Micro/Nanofluidic Devices for Single Cell Analysis, Volume II

Micro/Nanofluidic Devices for Single Cell Analysis, Volume II

Editors

Tuhin Subhra Santra
Fan-Gang Tseng

MDPI • Basel • Beijing • Wuhan • Barcelona • Belgrade • Manchester • Tokyo • Cluj • Tianjin

Editors
Tuhin Subhra Santra
Department of Engineering
Design
Indian Institute of Technology
Madras
Chennai
India

Fan-Gang Tseng
Department of Engineering
and System Science
National Tsing Hua University
Hsinchu
Taiwan

Editorial Office
MDPI
St. Alban-Anlage 66
4052 Basel, Switzerland

This is a reprint of articles from the Special Issue published online in the open access journal *Micromachines* (ISSN 2072-666X) (available at: www.mdpi.com/journal/micromachines/special_issues/single_cells_analysis_II).

For citation purposes, cite each article independently as indicated on the article page online and as indicated below:

LastName, A.A.; LastName, B.B.; LastName, C.C. Article Title. *Journal Name* **Year**, *Volume Number*, Page Range.

ISBN 978-3-0365-2919-6 (Hbk)
ISBN 978-3-0365-2918-9 (PDF)

© 2022 by the authors. Articles in this book are Open Access and distributed under the Creative Commons Attribution (CC BY) license, which allows users to download, copy and build upon published articles, as long as the author and publisher are properly credited, which ensures maximum dissemination and a wider impact of our publications.

The book as a whole is distributed by MDPI under the terms and conditions of the Creative Commons license CC BY-NC-ND.

Contents

About the Editors vii

Preface to "Micro/Nanofluidic Devices for Single Cell Analysis, Volume II" ix

Tuhin Subhra Santra and Fan-Gang Tseng
Editorial for the Special Issue on Micro/Nanofluidic Devices for Single Cell Analysis, Volume II
Reprinted from: *Micromachines* **2021**, *12*, 875, doi:10.3390/mi12080875 1

Axel Hochstetter
Lab-on-a-Chip Technologies for the Single Cell Level: Separation, Analysis, and Diagnostics
Reprinted from: *Micromachines* **2020**, *11*, 468, doi:10.3390/mi11050468 7

Shuya Sawa, Mitsuru Sentoku and Kenji Yasuda
On-Chip Multiple Particle Velocity and Size Measurement Using Single-Shot Two-Wavelength Differential Image Analysis
Reprinted from: *Micromachines* **2020**, *11*, 1011, doi:10.3390/mi11111011 41

Moeto Nagai, Keita Kato, Satoshi Soga, Tuhin Subhra Santra and Takayuki Shibata
Scalable Parallel Manipulation of Single Cells Using Micronozzle Array Integrated with Bidirectional Electrokinetic Pumps
Reprinted from: *Micromachines* **2020**, *11*, 442, doi:10.3390/mi11040442 57

Hongyan Liang, Yi Zhang, Deyong Chen, Huiwen Tan, Yu Zheng and Junbo Wang et al.
Characterization of Single-Nucleus Electrical Properties by Microfluidic Constriction Channel
Reprinted from: *Micromachines* **2019**, *10*, 740, doi:10.3390/mi10110740 71

Mingxin Xu, Wenwen Liu, Kun Zou, Song Wei, Xinri Zhang and Encheng Li et al.
Design and Clinical Application of an Integrated Microfluidic Device for Circulating Tumor Cells Isolation and Single-Cell Analysis
Reprinted from: *Micromachines* **2021**, *12*, 49, doi:10.3390/mi12010049 83

Fenfang Li, Igor Cima, Jess Honganh Vo, Min-Han Tan and Claus Dieter Ohl
Single Cell Hydrodynamic Stretching and Microsieve Filtration Reveal Genetic, Phenotypic and Treatment-Related Links to Cellular Deformability
Reprinted from: *Micromachines* **2020**, *11*, 486, doi:10.3390/mi11050486 95

Esra Sengul and Meltem Elitas
Single-Cell Mechanophenotyping in Microfluidics to Evaluate Behavior of U87 Glioma Cells
Reprinted from: *Micromachines* **2020**, *11*, 845, doi:10.3390/mi11090845 109

Meltem Elitas and Esra Sengul
Quantifying Heterogeneity According to Deformation of the U937 Monocytes and U937-Differentiated Macrophages Using 3D Carbon Dielectrophoresis in Microfluidics
Reprinted from: *Micromachines* **2020**, *11*, 576, doi:10.3390/mi11060576 123

Chen Zhu, Xi Luo, Wilfred Villariza Espulgar, Shohei Koyama, Atsushi Kumanogoh and Masato Saito et al.
Real-Time Monitoring and Detection of Single-Cell Level Cytokine Secretion Using LSPR Technology
Reprinted from: *Micromachines* **2020**, *11*, 107, doi:10.3390/mi11010107 139

About the Editors

Tuhin Subhra Santra

Dr. Tuhin Subhra Santra has been an Assistant Professor in the Department of Engineering Design at the Indian Institute of Technology Madras, India, since July 2016. Dr. Santra received his Ph.D. degree from National Tsing Hua University (NTHU), Taiwan. Dr. Santra was a Postdoctoral Researcher at the University of California, Los Angeles (UCLA), USA. His main research areas are Bio-NEMS, MEMS, single-cell technologies, single-molecule detection, biomedical micro-/nanodevices, nanomedicine, etc. He has served as a Guest Editor for *Frontiers of Bioengineering and Biotechnology, Cells* (MDPI), *International Journal of Molecular Sciences, Sensors, Molecules*, and *Micromachines*, among others. He was the conference chair and a committee member of IEEE-NEMS in 2017, 2020 and 2021. Dr. Santra has received many honors and awards such as "DBT/Wellcome Trust India Alliance Fellowship" in 2018, Honorary Research Fellow from National Tsing Hua University, Taiwan, in 2018, and IEEE-NEMS best conference paper award in 2014. He published more than 8 books, 40 SCI journals, 20 book chapters, 20 US/Taiwan/Indian patents, and 20 international conference proceedings in his research field.

Fan-Gang Tseng

Dr. Fan-Gang (Kevin) Tseng received his Ph.D. degree in Mechanical Engineering from UCLA, USA, in 1998. He joined Engineering and System Department of National Tsing Hua University in 1999 and advanced to Professor in 2006. He was the Chairman of the ESS Department at NTHU (2010–2013), a Visiting Scholar of Koch Institute of Integrated Cancer Research at MIT USA (2014–2015), and the Dean of Nuclear Science College at NTHU (2016–2017). His research interests are in the fields of BioNEMS, biosensors, microfluidics, tissue chips, and fuel cells. He has received 60 patents, written 8 book chapters, and published more than 260 SCI journal papers and 400 conference technical papers. He has received several awards, including Shakelton Scholar, National Innovation Award (twice), Outstanding in Research Award (twice), and Mr. Wu, Da-Yo Memorial Award from MOST, Taiwan, and more than 20 best papers and other awards in various international conferences and competitions. He is among the editorial board of several journals including *IJMS, Cells, Micromachines*, and *Applied Science*, and was also the general co-chair for MicroTas 2018 and a board member of CBMS from 2018 to 2022.

Preface to "Micro/Nanofluidic Devices for Single Cell Analysis, Volume II"

After over sixty years since Dr Feynman's awe-inspiring lecture at Caltech in the winter of 1959, our dependence on micro/nanotechnology has only been expanding. From ultra-efficient solar cells and high-power batteries to superfast computers, the scientific community is continually tuning different micro/nanoscale properties to make remarkable innovations in the area of functional high-speed electronics. In a diversifying digital era, where our dependence on fast communication and efficient energy consumption is increasing every day, micro/nanoscience is an integral part of how we are shaping our present while continuously laying the foundations for future technologies.

Apart from the developments in miniaturized electronics, another frontier in which micro/nanotechnology has been making significant progress is in the area of biological research. The ability to visualize and manipulate cellular activities at the nanoscale provides researchers with new insights into how different cellular functions occur. This may include understanding the different intracellular molecular behaviors, such as single-cell sequencing and gene expression, analyzing intracellular transformations and responses to external cargo, and understanding cell–environment interactions, which provide fundamental knowledge on cell proliferation and differentiation. An understanding of these biological properties at the single-cell level can help scientists to perform different omics analyses and help in studying different biological phenomena such as disease progression and tissue regeneration. The ultimate goal is to design strategies that are able to perform single-cell analyses in vivo. This includes the development of various lab-on-a-chip and organ-on-a-chip devices with necessary regulatory approval, which can be employed for rapid clinical applications.

The ability of micro/nanofluidic devices to precisely regulate the motion of biological samples, biomolecules, nanoparticles, and a single cell has come as a massive development in the area of biotechnology and biomedical engineering. These miniaturized devices are highly non-toxic, have very low manufacturing costs, and enable extremely delicate maneuvering of experimental parameters. As micro/nanochannels can hold samples at a measure as low as the picolitre scale, a high-throughput analysis of single cells can be performed. These specific advantages have greatly facilitated the investigation of intracellular behavior, extracellular interaction, and intercellular signaling. The past two decades have seen the scientific community utilize the benefits of micro/nanofluidic devices to make significant breakthroughs in understanding the fundamentals of cell behavior at the single-cell level. The potential for application in healthcare and diagnostics is immense.

This second edition of the Special Issue "Micro/nanofluidic devices for Single Cell Analysis" contains nine articles and one editorial, which explore the various avenues of single-cell analysis. The articles have been selected in such a way so as to put emphasis on the different single-cell technologies that have employed a micro/nanofluidic-based approach to perform intracellular detection, imaging, and monitoring. Some aspects of single-cell manipulation, analysis, characterization and/or diagnosis are also discussed. We hope that this Special Issue will be helpful in providing researchers with new insights into the field of single-cell analysis using micro/nanofluidic devices.

Tuhin Subhra Santra, Fan-Gang Tseng
Editors

Editorial

Editorial for the Special Issue on Micro/Nanofluidic Devices for Single Cell Analysis, Volume II

Tuhin Subhra Santra [1,*] and Fan-Gang Tseng [2]

1. Department of Engineering Design, Indian Institute of Technology, Chennai 600036, India
2. Department of Engineering and System Science, National Tsing Hua University, Hsinchu 300044, Taiwan; fangang@ess.nthu.edu.tw
* Correspondence: tuhin@iitm.ac.in

Citation: Santra, T.S.; Tseng, F.-G. Editorial for the Special Issue on Micro/Nanofluidic Devices for Single Cell Analysis, Volume II. *Micromachines* 2021, 12, 875. https://doi.org/10.3390/mi12080875

Received: 21 July 2021
Accepted: 23 July 2021
Published: 26 July 2021

Publisher's Note: MDPI stays neutral with regard to jurisdictional claims in published maps and institutional affiliations.

Copyright: © 2021 by the authors. Licensee MDPI, Basel, Switzerland. This article is an open access article distributed under the terms and conditions of the Creative Commons Attribution (CC BY) license (https://creativecommons.org/licenses/by/4.0/).

The functional, genetic, or compositional heterogeneity of healthy and diseased tissues promotes significant challenges to drug discovery and development [1,2]. Genetically identical cells may exhibit phenotype heterogeneity, which is of particular importance for tumor metastasis, stem cell differentiation, and drug resistance [3]. Such heterogeneities impede accurate disease modeling and can mislead the elucidation of biomarker levels, and may misguide patient responses to particular therapies [1–3]. Nevertheless, cellular heterogeneity has remained unexplored for a long time as former studies mainly focused on cell manipulation and analysis at the bulk scale, providing the average interpretation of the results. The complex nature of cells has been the long-standing motivation for developing the tools for single-cell transcriptomic, genomic, and multiplex proteomic analyses [1,4–6]. However, the traditional biological tools, including petri-dishes and well-plates, technically limit micron-scale single-cell manipulation and analysis. Additionally, the use of low concentrations target biomolecules introduces additional challenges in this field. Therefore, single-cell research provokes the access of modern technologies to address single-cell functionalities with high-throughput efficiency [1,4,7].

Single-cell technologies are beneficial for the studies of scarce cells [1,4,8,9]. For example, circulating tumor cells (CTCs) are rare, such as one in the background of 10^7 normal blood cells. Detecting and characterizing these cells could help and explore the underlying cause of cancer spreads and be very useful in developing efficient and targeted therapies [10,11]. Usually, single cells [1,4] have been isolated by multi-well plates in most biological labs, which provides low efficiency, and significant labor strength is needed. Another option is robotic liquid handling workstations that reduce labor intensity but are pretty expensive to install in the lab. The standard techniques, such as flow cytometry and laser scanning cytometry, can rapidly screen fluorescent-labeled cells in a flow, and these have been used for single-cell analysis for a long time [12,13]. Flow cytometry is an automatic technique for multiple detections and sorting of single cells. However, the instrument is expensive, bulky and mechanically complicated. It requires large sample volumes and analyzes cells at a one-time point. Hence, flow cytometry cannot provide continuous variation in the cell dynamics.

On the other hand, micro/nanofluidic devices have emerged as a potential platform with advanced technologies for single-cell manipulation and analysis in the last two decades. Micro/nanofluidic devices have many unrivaled advantages over conventional techniques [14–17]. They can manipulate and control fluids in the range of micro to pico-liters, thus reducing sample loss, providing susceptible analysis in the miniaturized microfluidic systems [14,15,17]. The micro/nanofluidic devices are not only designed and fabricated to fulfill the needs of various single-cell manipulation, separation, trapping isolation and lysis, but also used for electrical, mechanical, optical, biochemical characterization, as well as for therapeutic and diagnostic purposes [18–26]. Figure 1 shows worldwide microfluidic single-cell-related article publications in the last two decades, and it indicates the high demand of microfluidic devices for single-cell analysis and applications, evidenced by the increase in the number of scientific publications in each year. The micro/nanofluidic devices expedite remarkable high-throughput parallel manipulation and analysis of single cells, providing more accurate statistical results than bulk analysis and having a meaningful interpretation. Moreover, multifunctional devices can be integrated on the same chip to make it automatic, eliminating possibilities of contamination and error-free operations. Additionally, fluorometry, mass spectroscopy, and fluorescence microscopy can be integrated with microfluidic systems for achieving deeper insights into single-cell morphology and functionalities. Such steps can pave new avenues in this exciting field. This Special Issue of *Micromachines* entitled "Micro/nanofluidic devices for single-cell analysis" encompasses the recent advancements in single-cell analysis using micro/nanofluidic devices.

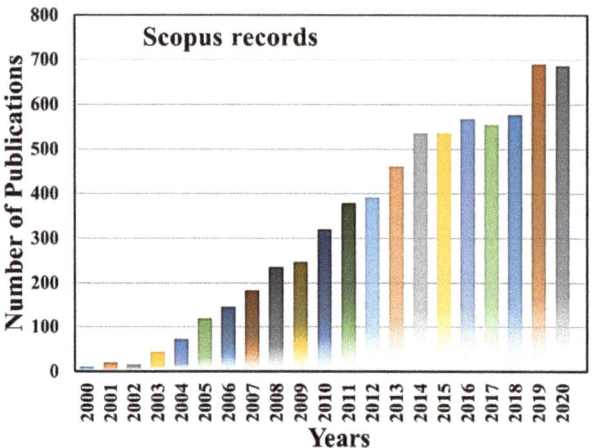

Figure 1. Year-wise microfluidic single cell scientific article publication. The data are adapted from Scopus records until 2020.

Hochstetter [27] briefly reviewed single-cell separation, diagnostics, and analysis using recent advancements in lab-on-a-chip technologies. Moreover, they reviewed the potentials, limitations, future prospects, and applications of microfluidic technologies, especially concerning the funding outlook and field requisition of the chips.

The measurement of sample flow velocity is essential for controlling the cell sorting time and reconstruction of image analysis. Sawa et al. [28] reported on-chip microparticle size and velocity assessment by using differential image analysis of single-shot two-wavelength. While the microparticles run via an image flow cytometer, they are irradiated by two different lights with different irradiation times simultaneously. For each wavelength of light, the images of the same microparticle were captured in a single shot. The velocity is calculated by comparing these two images: the difference of the particles' elongation divided by the difference of irradiation time. These accurate velocity and shape measurements can improve the cell sorter technique and the imaging flow cytometry to diagnose cells.

The high-throughput in vivo cellular microenvironment can permit us to investigate cellular function in detail. Nagai et al. [29] developed a parallel single-cell manipulation using a micronozzle array compacted with a bidirectional electrokinetic micropump. The polydimethylsiloxane (PDMS) micro nozzle array combined with bidirectional electrokinetic pumps are operated by using DC-biased AC voltages. Single HeLa cells were transported to the nozzle holes. After applying voltage, adequate electroosmotic flow occurs outside the nozzle array and manipulates the single-cell simultaneously.

Li et al. [30] demonstrated the hydrodynamic cell stretching and microsieve filtration, which can express the link between genetic, phenotypic, and treatment to the cellular deformability. This cellular deformability has shown the correlation between metastatic cancer cells and invasiveness. In-depth studies on single-cell hydrodynamic stretching can correlate mechanical characteristics of cancer cells with genetics and phenotypes. It helps to distinguish the differential deformability of cell models toward promoting drug treatment, EMT, and invasiveness, thus strengthening our knowledge on the fundamentals of cancer progression.

The electrical properties of the cells cannot be measured effectively by conventional techniques. Liang et al. [31] proposed a microfluidic-based constriction channel to characterize the electrical properties of a single nucleus. The device can isolate and trap single nuclei at the microfluidic channel without any pipette tips for electrical measurements. Their technique can classify cell type and cell status evaluation through bioelectrical markers of cell nuclei. Here, the authors studied the effect of membrane capacitance on the estimation of nuclear electrical properties and compared it with electrorotation.

Sengul and Elitas [32] presented sensitive, label-free, and specific, single-cell electrochemical properties using a microfluidic device. They fabricated a 3D carbon electrode array-based device and showed deformation measurement of U937 monocytes and dielectric movement and U937-differentiated macrophages in a less conductive medium. Using their technique, the cell damage caused by aggressive shear forces can be measured, and cells can be used for further downstream analysis. Moreover, these results also revealed that dielectric mobility and deformation could be exploited as an electromechanical biomarker to recognize differentiated cell populations from their progenitors.

The same group [33] investigated the impact of macrophages on glioma cell behavior by using a microfabricated cell culture platform. They quantified motility, migration, morphology, proliferation, and deformation characteristics of glioma U87 cells at the single-cell resolution to unveil biomechanical heterogeneity. They could quantify the mechanophenotypic properties of glioma cells by using their microfluidic device.

Cytokine secretion has a tremendous impact on clinical diagnostics. Zhu et al. [34] reported Cytokine secretion detection at the single-cell level and real-time monitoring using localized surface plasmon resonance (LSPR). The authors developed a microwell chip with cyclo-olefin-polymer (COP) film imprinted with gold-capped nanopillars for Interleukin 6 (IL-6) detection at the single-cell level. The trapped cell secret cytokine was analyzed by using the spectrum analyzer. This fabricated device facilitates real-time monitoring that can monitor the biological variation of the tested single-cell viability.

The CTCs can be considered a substitute approach for tissue biopsy, and it able to provide tumor-derived and germline-specific genetic variations. The analysis of the CTCs at a single-cell level can enable in-detail tumor heterogeneity exploration and individual clinical assessment. Xu et al. [35] demonstrated CTCs isolation and clinical application by using a microfluidic chip integrated with a micropore-arrayed filtration membrane. The device has the ability to provide CTCs isolation with high efficiency, throughput, and minimal damage of the cell. Moreover, the device can detect a positive detectable rate of 87.5% CTCs from lung cancer patients. This detection method can be a promising tool for cancer research and the accomplishment of CTCs analysis for routine clinical practice.

In conclusion, this Special Issue entitled "Micro/nanofluidic devices for single-cell analysis" not only covers single-cell manipulation, separation, diagnostics but also it discussed single-cell mechanical, electrical, and electrochemical characterizations and their analysis. Moreover, this Special Issue elaborates on cellular heterogeneity characteristics, Cytokine secretion detection, circulating tumor cell (CTCs) isolation, and clinical applications.

Author Contributions: T.S.S. wrote this editorial, and F.-G.T. corrected it. Both authors have read and agreed to the published version of the manuscript.

Funding: We acknowledge the DBT/Wellcome Trust India Alliance Fellowship for funding under grant number IA/E/16/1/503062.

Acknowledgments: The authors greatly appreciate Ashwini Shinde from the Department of Engineering Design, IIT Madras, Bio-µ-Nano Lab, for her help in writing this editorial and do the necessary corrections. We also acknowledge Srabani Kar from the Department of Electrical Engineering, the University of Cambridge, for the preparation of Scopus data and this editorial correction.

Conflicts of Interest: The authors declare no conflict of interest.

References

1. Santra, T.S.; Tseng, F.-G. (Eds.) *Handbook of Single-Cell Technologies*; Springer: Singapore, 2021; ISBN 978-981-10-8952-7.
2. Genetic effects on gene expression across human tissues. *Nature* **2017**, *550*, 204–213. [CrossRef]
3. Ackermann, M. A functional perspective on phenotypic heterogeneity in microorganisms. *Nat. Rev. Microbiol.* **2015**, *13*, 497–508. [CrossRef]
4. Tseng, F.-G.; Santra, T.S. *Essentials of Single-Cell Analysis: Concepts, Applications and Future Prospects*; MDPI: Basel, Switzerland, 2016; ISBN 978-3-662-49118-8.
5. Kanter, I.; Kalisky, T. Single Cell Transcriptomics: Methods and Applications. *Front. Oncol.* **2015**, *5*. [CrossRef]
6. Chappell, L.; Russell, A.J.C.; Voet, T. Single-Cell (Multi)omics Technologies. *Annu. Rev. Genom. Hum. Genet.* **2018**, *19*, 15–41. [CrossRef]
7. Shinde, P.; Mohan, L.; Kumar, A.; Dey, K.; Maddi, A.; Patananan, A.; Tseng, F.-G.; Chang, H.-Y.; Nagai, M.; Santra, T. Current Trends of Microfluidic Single-Cell Technologies. *Int. J. Mol. Sci.* **2018**, *19*, 3143. [CrossRef] [PubMed]
8. Santra, T.S.; Tseng, F.-G. (Eds.) *Single Cell Analysis*; MDPI: Basel, Switzerland, 2021; ISBN 978-3-0365-0629-6.
9. Santra, T.S.; Tseng, F.-G. Single-Cell Analysis. *Cells* **2021**, *9*, 1993. [CrossRef]
10. Castro-Giner, F.; Aceto, N. Tracking cancer progression: From circulating tumor cells to metastasis. *Genome Med.* **2020**, *12*, 31. [CrossRef] [PubMed]
11. Habli, Z.; AlChamaa, W.; Saab, R.; Kadara, H.; Khraiche, M.L. Circulating Tumor Cell Detection Technologies and Clinical Utility: Challenges and Opportunities. *Cancers* **2020**, *12*, 1930. [CrossRef] [PubMed]
12. Harnett, M.M. Laser scanning cytometry: Understanding the immune system in situ. *Nat. Rev. Immunol.* **2007**, *7*, 897–904. [CrossRef] [PubMed]
13. Brummelman, J.; Haftmann, C.; Núñez, N.G.; Alvisi, G.; Mazza, E.M.C.; Becher, B.; Lugli, E. Development, application and computational analysis of high-dimensional fluorescent antibody panels for single-cell flow cytometry. *Nat. Protoc.* **2019**, *14*, 1946–1969. [CrossRef]
14. Tseng, F.-G.; Santra, T.S. (Eds.) *Micro/Nano Fluidic Devices for Single Cell Analysis*, 1st ed.; MDPI: Basel, Switzerland, 2015; ISBN 978-3-03842-090-3.
15. Santra, T.S.; Tseng, F.G. Micro/nanofluidic devices for single cell analysis. *Micromachines* **2014**, *5*, 154. [CrossRef]
16. Kumar, A.; Shinde, P.; Mohan, L.; Mhapatra, P.S.; Santra, T.S. Microfluidic technologies for cell manipulation, therapeutics and analysis. In *Microfluidics and Bio-MEMS: Devices and Applications*; Santra, T.S., Ed.; Jenny Stanford Publisher: Singapore, 2020; p. 550, ISBN 978-1-003-01493-5.
17. Santra, T.S. (Ed.) *Microfluidics and Bio-MEMS: Devices and Applications*, 1st ed.; Jenny Stanford Publisher: Singapore, 2020; ISBN 9789814800853.
18. Shinde, P.; Kumar, A.; Illath, K.; Dey, K.; Mohan, L.; Kar, S.; Barik, T.K.; S-Rad, J.; Nagai, M.; Santra, T.S. Physical approaches for drug delivery—An overview. In *Delivery of Drugs*; Elsevier: Amsterdam, The Netherlands, 2020; Volume 2, ISBN 978-0-12-817776-1.
19. Santra, T.S.; Chen, C.-W.; Chang, H.-Y.; Tseng, F.-G. Dielectric passivation layer as a substratum on localized single-cell electroporation. *RSC Adv.* **2016**, *6*. [CrossRef]
20. Tseng, F.-G.; Santra, T.S. (Eds.) Electroporation for single cell analysis. In *Essentials of Single Cell Analysis*; Springer: Berlin, Germany, 2016; ISBN 978-3-662-49116-4.
21. Santra, T.S.; Kar, S.; Chang, H.-Y.; Tseng, F.-G. Nano-localized single-cell nano-electroporation. *Lab Chip* **2020**, *20*, 4194–4204. [CrossRef]
22. Manoj, H.; Gupta, P.; Mohan, L.; Nagai, M.; Wankhar, S.; Santra, T.S. Microneedles: Current trends and applications. In *Microfluidics and Bio-MEMS: Devices and Applications*; Santra, T.S., Ed.; Jenny Stanford Publisher: Singapore, 2020; ISBN 978-1-003-01493-5.
23. Kar, S.; Shinde, P.; Nagai, M.; Santra, T.S. Optical manipulation of cells. In *Microfluidics and Bio-MEMS: Devices and Applications*; Santra, T.S., Ed.; Jenny Stanford Publisher: Singapore, 2020; p. 550, ISBN 978-1-003-01493-5.
24. Kumar, A.; Mohan, L.; Shinde, P.; Chang, H.-Y.; Nagai, M.; Santra, T.S. Mechanoporation: Toward single cell approaches. In *Handbook of Single Cell Technologies*; Springer: Singapore, 2018; ISBN 978-981-10-8952-7.
25. Kaladharan, K.; Kumar, A.; Gupta, P.; Illath, K.; Santra, T.; Tseng, F.-G. Microfluidic Based Physical Approaches towards Single-Cell Intracellular Delivery and Analysis. *Micromachines* **2021**, *12*, 631. [CrossRef]
26. Kar, S.; Mohan, L.; Dey, K.; Shinde, P.; Chang, H.-Y.; Nagai, M.; Santra, T.S. Single Cell Electroporation-Current Trends, Applications and Future prospects. *J. Micromech. Microeng.* **2018**, *28*, 123002. [CrossRef]
27. Hochstetter, A. Lab-on-a-Chip Technologies for the Single Cell Level: Separation, Analysis, and Diagnostics. *Micromachines* **2020**, *11*, 468. [CrossRef]
28. Sawa, S.; Sentoku, M.; Yasuda, K. On-Chip Multiple Particle Velocity and Size Measurement Using Single-Shot Two-Wavelength Differential Image Analysis. *Micromachines* **2020**, *11*, 1011. [CrossRef]

29. Nagai, M.; Kato, K.; Soga, S.; Santra, T.S.; Shibata, T. Scalable Parallel Manipulation of Single Cells Using Micronozzle Array Integrated with Bidirectional Electrokinetic Pumps. *Micromachines* **2020**, *11*, 442. [CrossRef] [PubMed]
30. Li, F.; Cima, I.; Vo, J.H.; Tan, M.-H.; Ohl, C.D. Single Cell Hydrodynamic Stretching and Microsieve Filtration Reveal Genetic, Phenotypic and Treatment-Related Links to Cellular Deformability. *Micromachines* **2020**, *11*, 486. [CrossRef] [PubMed]
31. Liang, H.; Zhang, Y.; Chen, D.; Tan, H.; Zheng, Y.; Wang, J.; Chen, J. Characterization of Single-Nucleus Electrical Properties by Microfluidic Constriction Channel. *Micromachines* **2019**, *10*, 740. [CrossRef] [PubMed]
32. Elitas, M.; Sengul, E. Quantifying Heterogeneity According to Deformation of the U937 Monocytes and U937-Differentiated Macrophages Using 3D Carbon Dielectrophoresis in Microfluidics. *Micromachines* **2020**, *11*, 576. [CrossRef] [PubMed]
33. Sengul, E.; Elitas, M. Single-Cell Mechanophenotyping in Microfluidics to Evaluate Behavior of U87 Glioma Cells. *Micromachines* **2020**, *11*, 845. [CrossRef]
34. Zhu, C.; Luo, X.; Espulgar, W.V.; Koyama, S.; Kumanogoh, A.; Saito, M.; Takamatsu, H.; Tamiya, E. Real-Time Monitoring and Detection of Single-Cell Level Cytokine Secretion Using LSPR Technology. *Micromachines* **2020**, *11*, 107. [CrossRef] [PubMed]
35. Xu, M.; Liu, W.; Zou, K.; Wei, S.; Zhang, X.; Li, E.; Wang, Q. Design and Clinical Application of an Integrated Microfluidic Device for Circulating Tumor Cells Isolation and Single-Cell Analysis. *Micromachines* **2021**, *12*, 49. [CrossRef] [PubMed]

Review

Lab-on-a-Chip Technologies for the Single Cell Level: Separation, Analysis, and Diagnostics

Axel Hochstetter

Experimentalphysik, Universität des Saarlandes, D-66123 Saarbrücken, Germany; axel_hochstetter@web.de;
Tel.: +49-(0)681-302-2730

Received: 8 February 2020; Accepted: 25 April 2020; Published: 29 April 2020

Abstract: In the last three decades, microfluidics and its applications have been on an exponential rise, including approaches to isolate rare cells and diagnose diseases on the single-cell level. The techniques mentioned herein have already had significant impacts in our lives, from in-the-field diagnosis of disease and parasitic infections, through home fertility tests, to uncovering the interactions between SARS-CoV-2 and their host cells. This review gives an overview of the field in general and the most notable developments of the last five years, in three parts: 1. What can we detect? 2. Which detection technologies are used in which setting? 3. How do these techniques work? Finally, this review discusses potentials, shortfalls, and an outlook on future developments, especially in respect to the funding landscape and the field-application of these chips.

Keywords: microfluidics; single cell level; diagnostics; biomedical engineering; parasites; cancer; infectious diseases; point-of-care

1. Introduction

Since the advent of microfluidics 30 years ago, many applications have employed the advantages of microfluidic environments: small sample volumes, ready parallelization, high reproducibility, a vast span of experimental timescales, and a high control over local conditions (e.g., temperature, light exposure, flow velocity and direction, shear forces, diffusion, concentration gradients, viscosity, cell motility, etc.). One very prominent aspect of microfluidic research is diagnostics on the single-cell or even molecular level. The significance of recent research towards microfluidics-based single cell diagnostic chips is apparent in health care, in our homes, and also very prominently in the fight against the COVID 19 pandemic: The diagnostic targets range from circulating tumor cells (CTC) [1–6], over parasites in blood [7–15], male fertility [16–20], molecular markers for infections [11,15,21–23], cells of a specific stage in their life cycle [24,25], plant pathogens [26] and the SARS-CoV-2 proteome [27–32]. Depending on the exact target (either the entire cell or sub-cellular markers) there are different approaches to on-chip detection, each with their own underlying fundamentals and set of limitations. Additionally, there are additional synergetic possibilities (smartphones, optical traps, high throughput, personalized test, portability). This review presents a collection of the techniques proven useful for single cell diagnostic chips and explains the basic physical, chemical and biological effects that drive these technologies.

Due to the small volumes and the readily available lithographic procedures, diagnostic chips often are mass-producible, which makes individual tests cheap. This also opens up a great potential for portable diagnostics for global health issues [33] even in rural and remote locations, like endemic areas in Africa. Sadly, and very surprisingly, this potential often is not reached, and many sound diagnostic devices end in the "valley of death" [34,35]. In my opinion, this is due to a hole in the funding landscape; while many funding agencies (e.g., the Bill & Melinda Gates Foundation, UKRI, HFSP) offer grants to research the technologies and their application in a lab, the time required to actually adapt these technologies into field-applicable devices extends beyond the general timeframe of these grants and there are by far not enough grants that cover the adaption and deployment of field-ready devices. Furthermore, the adaptation of devices to make them field-applicable requires an inter- and cross-disciplinary skillset and contacts to both relevant populations and health-system officials. These requirements for device adaptation and deployment are massively at odds with how a research group has to be structured to be successful in today's academic and funding landscape. Sadly, Academia in most countries selects for high-throughput high-impact publishing researchers, while these Grand Challenges tend to be tackled by researchers who follow a conviction to better the world, and sacrifice their publication output along the way. If mankind really aims to not just face, but succeed in the face of these Grand Challenges, Academia needs to adapt, raise their standards beyond the number of the h-index.

What We Can Detect:

In recent years, more and more targets for detection on diagnostic chips have been investigated. Some salient examples thereof are circulating tumor cells (CTC) of various types of cancer [1,36–38], rare cells (e.g., sickle-cell variants of red blood cells) [39,40], parasites, like *Plasmodium falciparum* [1,7,10,11,13–15,21–23,36,41–50] and *Trypanosoma spp.* [8,44,51–57] and even plant pathogens [26,58], as well as—after cells have been lysed—subcellular infection markers (e.g., DNA, RNA fragments) [10,11,22,43,59–62]. Given the vast adaptability of microfluidics to any kind of single or multi-cellular assay [63], the ability to combine it with various light microscopy techniques [64], image processing [65], optical or acoustic traps [53], generation of chemical gradients [66], and even cell culture [4,67–83], any cellular or subcellular target seems to be possible for future on-chip diagnostics.

For easier access to the contents of this review, please find below a table which summarizes all the techniques discussed throughout this publication and their applications toward single cell diagnostic chips and beyond (see Table 1).

Table 1. Techniques applied to achieve single cell diagnostic chips.

Technique (Abbreviation)	Applications
Dielectrophoresis (DEP)	Separation of blood, infected blood cells, parasites
Deterministic lateral displacement (DLD)	Separation of blood, infected blood cells, parasites, CTC, spores, DNA, viruses
Deformability-based separation	Separation of blood, infected blood cells, parasites
Margination & Dean Flow	Separation of blood, infected blood cells, parasites, CTC
Surface acoustic waves (SAW)	Separation of blood, parasites, CTC, multicellular organism
Optical tweezers (OT)	Separation of rare cells, parasites, CTC
Optical density/refractive index	Fast optical analysis of cell size, shape and optic density
Droplet microfluidics	Separation and storage of cells, mixing with targeted start of reactions, post-lysis analysis of cell contents
Paper microfluidics	Separation of cells, post-lysis analysis of cell contents
PCR-based techniques	Targeted enrichment of subcellular fragments to verify their presence or identity of the previously separated cells
LAMP-based techniques	Similar to PCR, but generally also employable in environments with very limited resources
Prote-/metabol-/transcript-/polyomics	Gather data on all the parts of a cell, their transcriptomes or their metabolites respectively. Elucidating contents, pathways and interactions, (e.g., host–viral interactions of SARS-CoV-2).

2. Methodologies

For this review, all results in Google Scholar for literature related to "single cell diagnostic chips" and their permutations of the last five years were screened. Additionally, where technologies were mentioned, the relevant original research papers pertaining to these technologies were screened, cited, and summarized to yield a more rounded review. The findings are summarized below in three chapters:

Cell separation methods
Combined separation and analysis on chip
Molecular analysis of single cells

2.1. Cell Separation Methods

In the last three decades, since the advent of microfluidics, several methodologies have been developed to create single cell diagnostic chips. While these three paragraphs only list these methodologies, the next section of this publication describes each of these methodologies and the underlying principles or technologies. For separation of cells in a microfluidic setup, we can either work by encapsulating samples inside droplets, or by leaving the entire sample inside a continuous (aqueous) phase. If the sample is in the continuous phase, we can separate the target cells either using deterministic lateral displacement (DLD), ratchets, dean-flow, di-electrophoresis, surface acoustic waves (SAW), optical and acoustic tweezers or by using optical density/refractive index. In a continuous phase it is also possible to follow a temporal evolution along the flow of the device.

By introducing two immiscible phases (e.g., water and oil) it is possible to create droplets, inside which the sample can be portioned. Droplet microfluidics has so far been used together with di-electrophoresis, deterministic lateral displacement, surface acoustic waves, or hydrodynamic droplet sorting to separate cells. It also has great potential, because it can be used to generate vast libraries of individual reaction volumes. This makes it possible to screen how identical cells react to entire sets of parameters in one single setup [84,85]. This can also be used to parallelize the analysis or separation of several sets of samples.

In case the target is not the entire cell, but a sub-cellular marker (e.g., DNA or RNA, individual antibodies [62] or other molecules) the cells containing the target molecule are lysed and/or the non-relevant cells are separated. After lysing the cells, their contents are optionally selectively multiplied (using e.g., polymerase chain reaction, PCR, and related techniques, like qPCR and nested PCR or Loop-mediated Isothermal Amplification, LAMP, and its derivatives like NINA-LAMP and LAMPport) and analyzed using chromatographic approaches (e.g., metabolomics, transcriptomics, proteomics, polyomics) or antibody-antigen binding, especially in rapid diagnostic tests (RDTs) [62]. This can all be done either in continuous phase microfluidics, droplet microfluidics, or paper-based microfluidics.

2.1.1. Dielectrophoresis (DEP)

Dielectrophoresis (DEP) describes the movements of cells (and other dielectric particles) within in a non-uniform electric field. This movement is caused by the induction of an electric dipole moment on the cell or particle, and the force the electric field gradient exerts on this dipole moment. It is not necessary that a cell carries a net surface electric charge to be polarized [8]. The induced dipole is aligned along the gradient of the non-uniform electric field. Thus, the coulomb forces generated on both sides of the dipole/particle are different, and a net force drags the particle across the gradient. The direction and strength of this force depends on many factors, both of the particle itself and the surrounding medium including the particles exact shape [86] and the frequency of the alternating current (AC) electric field [54]. For more details, see a recent review by Adekanmbi and Srivastava [86].

In short: by varying the frequencies of the employed AC electric field and/or the conductivity of the surrounding medium (e.g., by adding salts to increase the medium's conductivity), different cells can be separated. A prominent example is the separation of red blood cells (RBCs) and trypanosomes (a unicellular pathogenic parasite), conducted by Menachery et al. [54]. They used a micro-fabricated gold four-arm spiral quadrupole electrode array (see Figures 1 and 2), with each arm arranged at 90° to each other and separated by 400 µm operated at frequencies ranging from 10 kHz to 400 kHz and with solution conductivities varying from 16 to 60 mS/m.

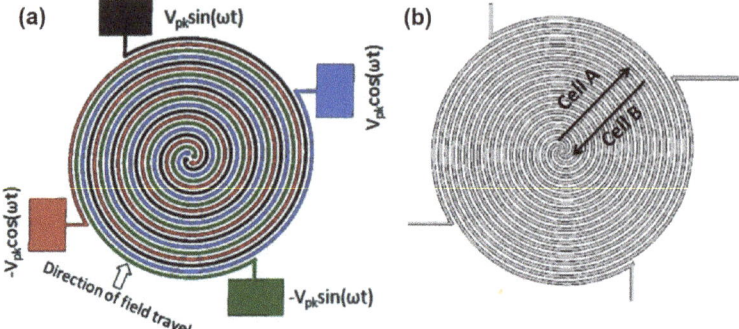

Figure 1. Four-arm spiral quadrupole electrode used by Menachery et al. [54]. (**a**) Schematic of the four-arm spiral microelectrode array comprising four parallel spiral elements of 30 mm in width and spacing. The electrodes are energized with a 90° phase shift with respect to each other. (**b**) Working principle of the chip. While cell type A (e.g., red blood cells) is expelled from the electrode array, cell type B (e.g., trypanosomes) is concentrated into the center of the array. Both processes take place simultaneously. Reproduced with permission from [54].

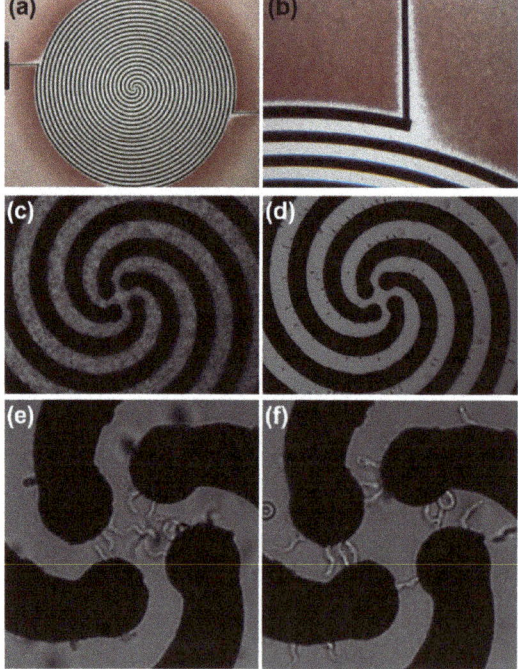

Figure 2. Enrichment of trypanosomes from infected blood. Total width of the spiral array is 2.9 mm, electrode width and spacing is 30 mm. (**a**,**b**) Micrograph following a separation process, with the RBCs having been pushed away from the electrode array. (**c**) Parasitized blood on the spiral electrode array. (**d**) Mouse RBCs are levitated and carried to the outer edges of the spiral. (**e**) Trypanosomes accumulate in the center of the spiral and undergo circular translational motion. (**f**) Trypanosomes are trapped along the electrode edges in the center of the spiral upon switching the AC voltage from quadrature-phase to an opposing two-phase. Reproduced with permission from [54].

Within this setup, it was possible to separate trypanosomes from murine RBCs at 140 kHz, and from human RBCs at 100 kHz and a Voltage of 2 V peak-to-peak, respectively [54]. This demonstrates that it is possible to completely separate different cell types from the same sample, simply based on their induced dipole moment. Since the induced dipole moment is specific for healthy cells (e.g., RBCs), infected cells (e.g., RBCs infected by *Plasmodium falciparum*), and pathogenic cells (e.g., trypanosomes) they can be separated in such an experimental setup [6,8,54,86,87] (see Figure 2). In many cases, the detection of a single infected cell or individual pathogen within a real-life sample can arguably be counted as a diagnosis of the respective infection (here: Malaria and Trypanosomiasis, respectively).

DEP can also be combined with other technologies for single cell diagnostic chips. One example is the combination of DEP with deterministic lateral displacement (DLD) as demonstrated by Jason Beech and others [6,8,88–90]; another prime example is the combination with SAW (see Section 2.1.5.), as used by Smith et al. to extract viable mesenchymal stroma cells from human dental pulp [91]. A drastically different usage of DEP is shown by Noghabi et al., who developed a same-single-cell analytic DEP chip, to study multidrug resistance inhibition in leukemic cells [92]. To achieve higher throughput, Faraghat et al. used three-dimensional DEP electrodes featuring tunnels, along which the cells were separated in a more continuous fashion [93].

For an overview of the pros and cons of DEP in the context of single cell separation and diagnostics, refer to Table 2.

Table 2. Pros and cons of dielectrophoresis (DEP).

Pros	Cons
Adaptable to different cells by frequency	Needs electricity
Potentially parallelizable	Low throughput, unless 3D- electrodes are used
Compatible with other techniques (e.g., DLD, SAW molecular analysis after cell lysis)	Cell frequencies need to be experimentally found
-	Separation is rather slow and hard to automate

2.1.2. Deterministic Lateral Displacement (DLD)

One core microfluidic technique used to separate cells by their size and shape (or rather, their effective hydrodynamic diameter) is deterministic lateral displacement (DLD), which was discovered accidently by Huang in 2004 [94]. There, laminar flow through a repetitive array of obstacles resulted in an asymmetrically bifurcated flow. This asymmetric flow separated different particles according to their diameter. In the 15 years since this empirical description of a phenomenon, additional aspects of DLD have been developed and have broadened its applications: the continuous separation of particles, cancer [3,95,96], healthy blood cells [40,97], infected RBCs [98], yeast [99], bacteria [24], fungal spores [100] and subcellular particles like DNA [101], and virus capsids [102]. Additionally, antibody-coated DLD arrays have been used for non-invasive prenatal diagnosis of circulating fetal cells in samples of their mother's blood [60]. Similarly, Hou et al. reported an antibody-coated "nanoVelcro" assay that selectively retains circulating fetal nucleated cells from blood samples of pregnant women [103].

The basic model that is used to describe how DLD works, is referred to by experts as a "naïve model", as it does not fully represent the physics behind the process, but helps to understand the separation that occurs, on a superficial level. This "naïve model" is based on dividing the flow through the DLD array into separate streams. The number of streams depends on the geometry of the DLD array (see Figure 3). The array is often made of rows of pillars that are shifted by a fraction 1/N of the row's width (which equals to the diameter of the pillar and the gap between two adjacent pillars). Thus, every N rows, the position of the pillars is the same (see Figure 3), and the fluid flow is divided into N streams. This row shift of 1/N is also denoted as row shift ε. Each stream carries the same current of fluid. Since the flow speeds vary across the gap and in between the rows, the streams are not of the same width. Especially around the pillars the streams are especially narrow. If a particle or cell has a diameter bigger than these narrow streams, they are—in the "naïve model"—not able to follow the stream and migrate to the next stream, while a smaller particle can follow the streamline along the flow (see Figure 3a). In this way, bigger particles get "bumped" along the array towards the side (perpendicular to the flow), and separated from smaller ones, that can "zig-zag" between the pillars and follow the flow.

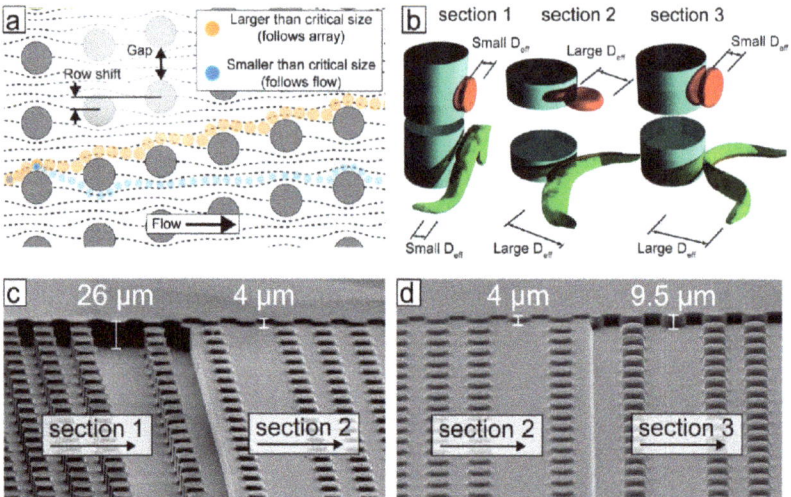

Figure 3. The principle the "naïve model" of deterministic lateral displacement, on the example of trypanosomes and red blood cells. (**a**) An array of posts divides a fluid-flow into many well-defined streams. Particles smaller than the critical size follow the streams, whereas larger particles follow a trajectory defined by the geometry of the array. (**b**) The effective size of particles is a function of their shape and orientation as they flow through the device. Device depth can be used to control the orientation of blood cells and parasites to maximize differences in effective sizes. (**c**,**d**) Scanning electron microscopy images of a poly(dimethylsiloxane) (PDMS) device, designed to separate trypanosomes from blood cells; the different sections achieve different separation steps. Adapted with permission from [8].

A more detailed discussion on more accurate models (including multiple critical diameters [104] and particle–particle interactions [105,106]) will be published soon by this author.

The shape of a cell or particle in a DLD array can have a massive impact on its hydrodynamic radius. The curved and elongated shape of trypanosomes and the biconcave shape of an RBC might inhibit a proper separation if the height of the DLD array is not carefully selected (see Figure 3b). Meanwhile, with an overly high array both cell types—despite being massively different—exhibit a very similar effective hydrodynamic diameter, simply because the trypanosomes can undulate freely (see Figure 3b). If the array is chosen to be very shallow (e.g., 4 μm), both cell types are forced to pass through the arrays while being vertically constricted. With an intermediate height, however, the larger trypanosomes are forced to undulate horizontally, while RBCs can align themselves vertically, resulting in a pronounced difference of their respective hydrodynamic radius.

NB: The complex motility of trypanosome has here been described as "undulating" for reasons of simplicity. A more detailed model of its motility has been discussed by Alizadehrad et al. in 2015 [107].

One of the disadvantages of DLD is that the arrays need to be tailored towards the cells which are to be separated and diagnosed on the chip. In addition, this tailoring is done by adapting the geometry of the DLD array: size and shape of the pillars, how far these pillars are apart, and the angle between the pillars and the main channel walls/the row shift ε. However, there have been several approaches to make DLD arrays "tunable" and thus adaptable over a larger range. One way to tune the sizes of separated cells is the combination of DEP and DLD [88–90]. Another way of tuning is combining DLD with non-Newtonian fluids, which change their viscosity depending of the flow velocity (and thus the shear forces the pillars exert on the liquid; shear-thinning) [108]. For general combinations of passive separation techniques (e.g., DLD) with active separation techniques (e.g., DEP, SAW, optical or magnetic) Yan et al. coined the phrase "hybrid microfluidics" [109].

For an overview of the pros and cons of DLD in the context of single cell separation and diagnostics, refer to Table 3 below.

Table 3. Pros and cons of deterministic lateral displacement (DLD).

Pros	Cons
Needs no electricity (e.g., using handheld syringes)	Prone to clogging
Tiny defects can be tolerated by redundancy	Trapped air bubbles can completely ruin separation
Can be parallelized	Low throughput due to tiny volume
Can be run constantly and separated cells can be collected for further processing in closed loop systems.	The separation process of DLD has not been fully understood so far and entails many different aspects that are thus far neglected.

Note: while the geometry of the array needs to be tailored to the cells that are to be separated in bulk samples, several techniques have been discovered that can tune the effective separation diameter to adapt an existing array to new cell types.

Additionally, many DLD applications work based on a naïve and simplified model.

2.1.3. Deformability-Based

A further technique for cell separation towards diagnostic chips that has been employed—both within and without DLD—is deformability-based separation. Not only are different cell types of different stiffness or elasticity (often measured in the Young's modulus), but also the infection with parasites can alter the elasticity of cells. The most prominent example thereof is the increased stiffness of red blood cells infected with *Plasmodium falciparum* (iRBCs) compared to healthy red blood cells (RBCs) [110]. While this fact in itself has intriguing implications with sickle-cell anemia [110] and immunity towards Malaria, it has also been used to separate iRBCs from RBCs, both in theory [98,110,111] and practice [14,112]. One especially clever approach was presented by Guo et al. in 2016, where an oscillating flow separated rigid from elastic RBCs through an asymmetric filter array (see Figure 4).

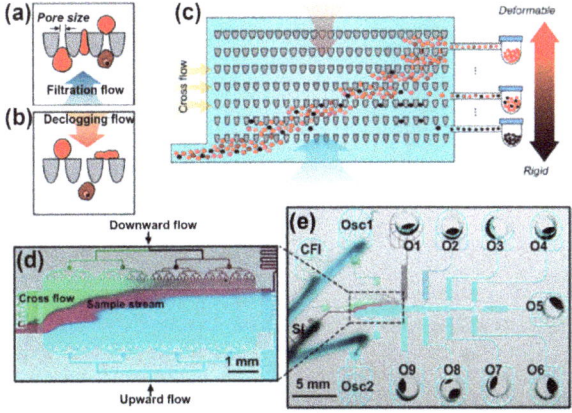

Figure 4. Design of the ratchet-sorting device. (**a**,**b**) Tapered funnel constriction allowing unidirectional flow of cells under oscillation excitation which consists of (**a**) upward filtration flow and (**b**) downward de-clogging flow; (**c**) cell sorting using a matrix of funnel constrictions. The cell sample is introduced

through the sample inlet (SI) and forms a diagonal trajectory under the combined forces of cross-flow inlet (CFI) and biased-oscillation flows including oscillation inlet 1 (Osc1) for de-clogging and oscillation inlet 2 (Osc2) for filtration. More deformable cells, such as RBCs, will travel further up the matrix of funnel constrictions than less deformable cells, such as iRBCs, which will be blocked midways and be separated from the main population. (**d**) Image of microfluidic ratchet device infused with different food color dyes illustrating the diagonal trajectory of the SI through the ratchet-sorting device constituting a deformability gradient. (**e**) Image of the overview design of the ratchet sorting device as well the nine outlets (O1–9). Adapted with permission from [112].

In the same year, Park et al. improved this approach to separate white blood cells (WBCs), RBCs and CTCs from each other [113]. Wang and coworkers recently combined deformability-based separation with magnetic-based techniques to separate CTCs, RBCs, and WBCs from "liquid biopsy" samples [114]. In 2018, Hongmei Chen reported the separation of CTCs, RBCs and WBCs from spiked peripheral blood samples using a combination of deformability-based and inertial separation [115]. Zhou et al. predicted that it is possible to combine deformability-based and electrokinetic separation, which relies on different shear moduli instead of ratchets [116]. While their computations are based on inanimate particles, this could lead to continuous cell and particle separation in ratchet-free and clogging-resistant devices.

For an overview of the pros and cons of deformability-based separation assays in the context of single cell level diagnostics, kindly refer to Table 4.

Table 4. Pros and cons of deformability-based assays.

Pros	Cons
Generally reusable setups	Defects in the matrix can undo separation
More resistant to clogging than DLD	Trapped air bubbles can trap deformable cells
Can theoretically be parallelized	Low throughput due to tiny volumes
Can run even without ratchets	Cannot be run constantly (due to oscillation), unless paired with other approaches e.g., DLD, immuno-magnetic, inertial, or electrokinetic sorting

2.1.4. Margination and Dean-Flow

Furthermore, independently of a DLD array, iRBCs have been separated from healthy RBCs using a constriction within a microfluidic channel at high hematocrit values (=high concentration of RBCs in the sample, basically an only slightly diluted blood sample) [117]. This separation effect, in which some cells (e.g., iRBCs, leukocytes) are accumulated along the margins of a long channel has been called margination [117,118] and is a naturally occurring phenomenon in our smaller blood vessels [7]. It can be even enhanced by using a viscoelastic fluid as medium to create a high-throughput detection system [13]. Similarly, spiral channels have been used to separate different cells by creating a Dean flow.

Xiang et al. recently even combined a Dean drag force-inducing spiral channel with a DLD array for a two-step separation of CTC from RBCs and white blood cells (WBCs) [119]. The underlying concept is that within curved channels two opposing forces are active on all particles and cells, the Inertial Lift force F_L and the Dean drag Force F_D.

The Inertial Lift force F_L is composed of the shear-gradient-induced lift (caused by the parabolic flow profile inside the curved rectangular microchannel) and the wall effect (caused by the asymmetric wake of the particle near the wall), which push neutrally buoyant particle away from the center of the channel and the walls, respectively. This force F_L can be calculated by [120–124]:

$$F_L = f_L(Re, x_L) \rho v_{max}^2 \cdot 16 r^4 / D_h^2 \qquad (1)$$

where the lift coefficient f_L is a function of the Reynolds number of the flow Re ($Re = \rho v_{max} D_h/\mu$) and the particle position x_L along the channel's cross-section (in respect to the channel's center). ρ and μ are the density and the dynamic viscosity of the fluid, respectively. v_{max} denotes the maximal velocity within the microchannel, D_h is the hydrodynamic diameter of the microchannel and shape and aspect ratio dependent (in first approximation: $2 \cdot \frac{Width \cdot Height}{Width + Height}$), and r is the particle radius.

The Dean (drag) force F_D on the other hand, will only form at relatively high Reynolds numbers in curved microchannels, due to the non-uniform inertia of the fluid in the inner and outer segments [124] of the channel. This Dean flow consists of two counter-rotating Dean vortices forming in the top and bottom halves of the channel (see Figure 5), and exerts additional transverse drag forces on particles [124].

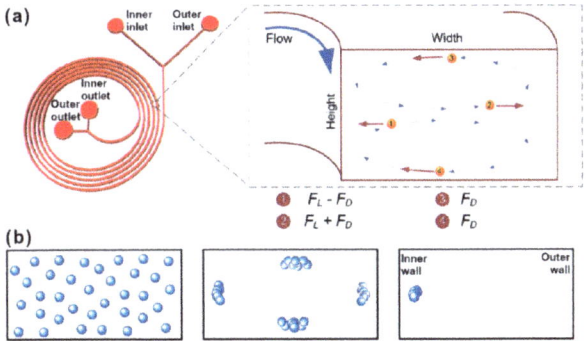

Figure 5. (a) Schematic of a spiral micro-particle separator. The design consists of two inlets and two outlets, with the sample being introduced through the inner inlet. Neutrally buoyant particles experience Lift forces (F_L) and Dean drag (F_D), which results in differential particle migration within the microchannel. (b) Microchannel cross-sections illustrating the principle of inertial migration for particles with $r/D_h \sim 0.05$. The randomly dispersed particles align in the four equilibrium positions within the microchannel where the Lift forces balance each other. Additional forces due to the Dean vortices reduce the four equilibrium positions to just one near the inner microchannel wall. Adapted with permission from [125].

This results in a displacement of particles flowing in a curved microchannel at high velocities. The flow in such a curved channel is defined by the dimensionless Dean number (De) [126]:

$$De = Re \sqrt{\frac{D_h}{2R}} \qquad (2)$$

with R as the radius of the channel's curvature. The maximum Dean drag force acting on a particle or cell in a curved channel can be estimated by Stokes drag force [125–127]:

$$F_D = 6\pi\mu v_{max} r = 1.08 \cdot 10^{-3} \cdot \pi\mu r \cdot De^{1.63} \qquad (3)$$

As can be seen from Equations (1) and (3), these two forces scale very differently with the radius of the particle (linear vs. the power of four), hence it is possible to separate different cells (or other particles) by their size at higher flow velocities in a curved rectangular microfluidic channel.

Further applications for Dean-flow-based microfluidics include the enrichment of human breast cancer cells from samples [128], human prostate epithelial tumor cells [129], or pathogenic bacteria from diluted blood samples [130]. To improve the cell separation prowess of Dean-flow-based applications, there have been some interesting recent improvements: through the addition of microstructures inside the channels, it is possible to separate blood cells using straight channels, which reduces the footprint of the designs and allows parallelization, as shown by Wu et al. in 2016 [131].

For an overview of the pros and cons of Dean-flow usage in the context of single cell separation and diagnostics, refer to Table 5.

Table 5. Pros and cons of Dean-flow-based approaches.

Pros	Cons
Generally reusable setups	Comparably big footprint
Can theoretically be parallelized	Prone to clogging
Does not need high resolution lithography	
Can handle higher cell densities	-
Bigger volumes can be handled	
Can be run continuously	
Can be combined with other technologies (e.g., droplet MF)	-

2.1.5. Surface Acoustic Waves (SAW)

Another technology that recently received increased attention in combination with microfluidic single cell diagnostics is called surface acoustic waves (SAW). SAW are generated by applying two conductive interdigital transducers (IDT) onto a piezo-electric carrier material. Interdigital here is rooted in Greek, and translates to "between fingers", meaning that two transducers are shaped like combs that are pushed into each other, without actually touching (see Figure 6).

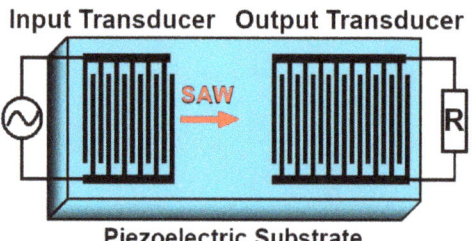

Figure 6. Interdigital transducers (IDT) (black) used to generate surface acoustic waves (SAW) on a piezoelectric carrier material. The input transducer generates mechanical forces from electrical voltage, while the output transducer can re-convert the mechanical oscillations into an alternating voltage.

If a voltage is applied between these comb-shaped transducers, then the piezoelectric material in between is forced to dilate or contract. Applying an alternating voltage forces the piezoelectric material into oscillations, which then are called surface acoustic waves. These surface acoustic waves can either be standing (SSAW) or travelling (TSAW), depending on the design of the IDT and the frequencies of the alternating voltage. Now, a microfluidic channel can be superimposed onto this piezoelectric substrate and carry cells and particles perpendicularly to these surface acoustic waves (see Figure 7).

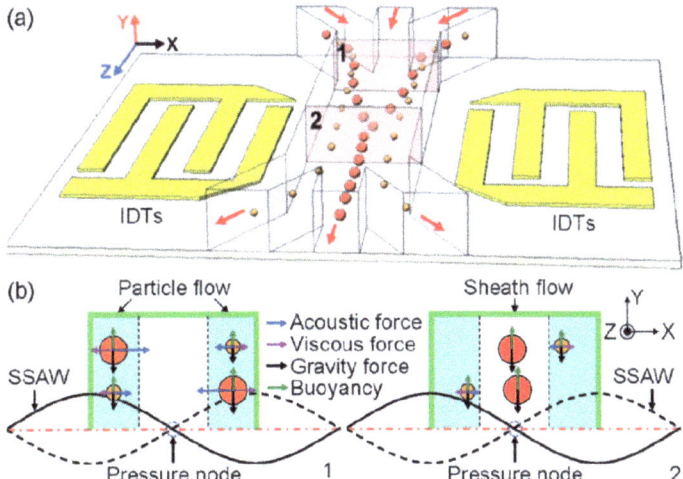

Figure 7. (a) Schematic of the separation mechanism showing particles beginning to translate from the sidewall to the center of the channel due to axial acoustic forces applied to the particles when they enter the working region of the SSAW (site 1). The differing acoustic forces cause differing displacements, repositioning larger particles closer to the channel center and smaller particles farther from the center (site 2). (b) Comparison of forces (normally in the pN range) acting on particles at site 1 and site 2, respectively. Reproduced with permission [132].

SSAW exert acoustic forces (F_a) on these cells and particles proportional to their volume [132]:

$$F_a = -\left(\frac{\pi p_0^2 V_p \beta_m}{2\lambda}\right) \phi(\beta, \rho) \sin 2kx \tag{4}$$

$$\phi = \frac{5\rho_p - 2\rho_m}{2\rho_p + \rho_m} - \frac{\beta_p}{\beta_m} \tag{5}$$

with p_0 as the pressure amplitude caused by the SSAW, V_p the volume of the particle, $\beta_{m,p}$ the compressibility of the medium and particle, respectively, λ for the ultrasonic wavelength, $\rho_{m,p}$ for the density of the medium and the particle, k for the wave vector and x for the distance to the pressure node.

Along the same axis as F_a, but with opposite direction, there are viscous forces F_v, which scale with the radius of the particle:

$$F_v = -6\pi \eta r v \tag{6}$$

with η for the medium viscosity, r for the particle radius and v the relative velocity of the particle with respect to the medium. Hence, acoustic forces dominate in larger particles but not in smaller particles, which allows the usage of SSAW to separate particles by size, if the other parameters are chosen accordingly.

TSAW can also be used for particle separation, but the acoustic radiation forces on cells and particles are much lower compared to SSAW at the same frequency and power input [133]. While higher actuation frequencies can increase the effectiveness of TSAW, they also can disturb the laminar flow within the microfluidic device by causing acoustic streaming [134,135]. This in itself, however, has given rise to other useful applications of TSAW [133–135], including pumping [136] and processing of multicellular organisms (e.g., *Caenorhabditis elegans*) [137].

A set of four IDTs, arranged along the sides of a square, can be used to create so called acoustic tweezers, where the interferences of the four IDTs can create trapping nodes that can hold individual cells. By changing the frequencies and amplitudes of the SAWs emitted by the IDTs, the trapped cells can be moved within the volume of the acoustic tweezers [138]. Similar to SAW in general, there can be three types of acoustic tweezers: (1) Standing-wave tweezers, (2) Traveling-wave tweezers, and (3) Acoustic-streaming tweezers [138].

For an overview of the pros and cons of SAW technology in the context of single cell separation and diagnostics, please refer to Table 6 below.

Table 6. Pros and cons of SAW-based approaches.

Pros	Cons
Tunable to different cells by changing frequency	Needs electricity
Work also with bigger, multicellular organisms	Needs piezoelectric and other expensive material
Can be parallelized	Calculations of optimal settings for acoustic forces not yet fully elucidated
Can be run continuously	Small volumes
Can be combined with other techniques	Forces might not fully compete with motile cells

2.1.6. Optical Trap and Tweezers

Light can interact with matter on the microscopic scale, as proven by Arthur Ashkin almost five decades ago [139]. The operating principle of optical tweezers is rather simple (see Figure 8), as demonstrated on laser light (orange) and a transparent spherical particle. A normal laser beam has a roughly Gaussian intensity profile (see orange intensity profile top left in Figure 8) and parallel rays of photons. If these rays enter a spherical particle of higher refractive index (e.g., a glass microsphere) the photons get refracted and in turn exert a force onto the sphere. Since there are more photons per interval where the intensity is higher, the sphere will pulled towards the center (by the gradient force $F_{gradient}$, see Figure 8) and pushed along the propagation of the laser beam (by the force of the scattered photons $F_{scatter}$).

If an additional lens is introduced into the light path, a position is created where the sphere can be held stably, since all the forces acting onto the sphere will be in equilibrium (see Figure 8 right). If the particle moves out of this point, the sum of the resulting forces will pull the particle right back into this stable position. Additionally, if the beam is moved away, the particle will be forced to follow. This is the basic on which optical trapping operates.

Since many cells have a higher refractive index (or are optically denser) than water or their culture media, they can be manipulated with optical traps, akin to the particle in Figure 8. Objects that have a lower refractive index than the surrounding medium (e.g., air bubbles), will get pushed away from the beam.

For alternative setups of optical traps, it is also possible to use two counter-propagating beams [140,141], or two parallel lasers to generate interference to trap particles [142], or two inclined fiber-coupled laser beams, which create a stable trapping position below the intersections [143].

Another way to generate optical traps is by shining an expanded laser beam onto a spatial light modulator (SLM) [144–150]. The resulting optical traps are generally referred to as "holographic optical traps (HOTs)" and are based on the interferences of many parallel phase shifts, all caused by being reflected off the SLM. The advantage of HOTs is that with one setup it is easy to generate dozens of traps, and move them in all 3 dimensions. An intermittent problem—the emergence of unwanted additional HOTs (so called ghost traps [151]) has in the meantime been reduced by the introducing small disorder [152], or random phase elements [153].

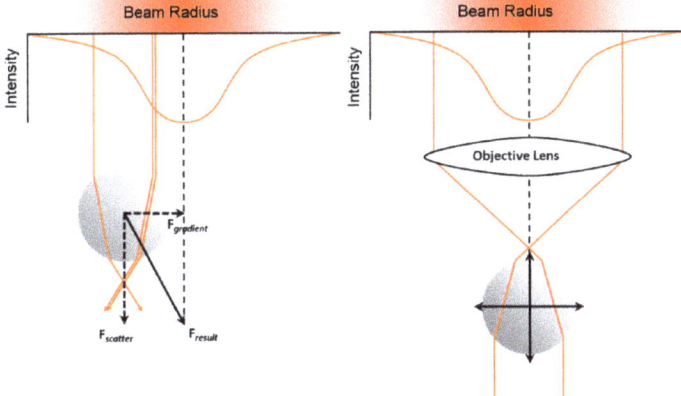

Figure 8. Operating principle of an optical trap/optical tweezers, demonstrated on laser light (orange) and a transparent spherical particle. (**Left**) A normal laser beam has a roughly Gaussian intensity profile (see orange intensity profile top left) and parallel rays of photons. Spherical particles of higher refractive index (e.g., a glass microsphere) refract photons that pass through them. By the law of impulse conservation, a force is exerted on the sphere (the scatter force $F_{scatter}$), which propels the particle along the propagation direction of the laser. The higher intensity of photons at the center of the beam, results in a net force towards the center of the beam (by the gradient force $F_{gradient}$). (**Right**) The introduction of an additional lens into the light path, creates stable trapping position. Here all optical forces acting onto the sphere are in equilibrium. Dislocating the particle from this point, results in a force that will pull the particle right back into equilibrium position. Image adapted from a sketch by Dr Eric Stellamanns.

The application of optical tweezers for the handling of single cells and for diagnostics on the single cell level has recently been discussed a lot [138,141,154,155]. However, it can as yet not be used as the only technique for any kind of diagnostics, and is either combined with other tests to create assays [63,66], or needs massive automation and robotics [156] or artificial intelligence to arrive at a diagnosis [156,157]. The usage of optical traps for single cell diagnosis is mostly limited to optical stretching [158], cell-sorting [66,154,159–162], and measuring of refractive indices [140,162–164]. Especially in combination with automatization, optical traps for single cell diagnostics have a great potential, but currently it is among the more expensive and less-performing technologies.

The pros and cons of optical traps in the context of single cell separation, analysis and diagnostics are listed in Table 7 below.

Table 7. Pros and cons of optical tweezer approaches.

Pros	Cons
Contact-free micromanipulation	Needs electricity and lasers & optical setup
Works intracellularly	Relies on differences in refractive indices (e.g., DNA-rich parts like the nucleus)
Can be used to measure forces on the cellular level	Potential photodamage to sample and devices at wrong wavelengths and higher exposures
Can be combined with other techniques	Forces might not fully compete with motile cells
Strong local forces can be achieved	However, only in "transparent" samples
An SLM-setup allows cell manipulation in 3D	Hard and expensive to parallelize

2.2. Combined Separation and Analysis on Chip

2.2.1. Droplet Microfluidics

All the technologies discussed so far have two things in common: they are all only used to separate different kinds of cells; and they all are used in continuous phase microfluidics, which means that one stream of liquid contains all the cells and particles. In contrast, there exist several techniques for which the cells are encapsulated in individual droplets (mostly of water or aqueous solutions) that are separated by another phase, either oil or even air. This allows not only to encapsulate single cells within individual droplets, but also to analyze their secretions, metabolites and (after cell lysis) their contents. Kaminski and Garstecki published a comprehensive overview on those techniques in 2017 [165]. In brief, there are the following techniques:

Droplet microfluidics: Any emulsion system of water droplets in an oil carrier phase (or water-droplets-in-oil-droplets-in-water; or any permutation thereof) can be used to separate individual cells or solutions from each other. This is generally done with an x-intersection on a microfluidic chip, where the aqueous phase is pinched off into droplets by two streams of oil (see Figure 9a) Using any kind of forces (shear, drag, coulomb, inertial, etc.), droplets can be merged to combine their contents and to start reactions towards a diagnostic signal (see Figure 9c), like for example an exosome immunoassay for cancer diagnosis [166].

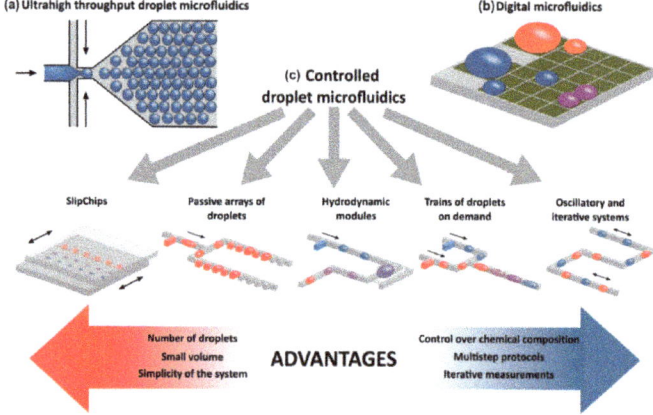

Figure 9. Schematic classification of droplet microfluidics that comprises three main groups of technologies: (**a**) ultrahigh throughput microfluidics characterized by the largest number of droplet bioreactors, (**b**) digital microfluidics that enables individual control over each droplet bioreactor, and (**c**) controlled droplet microfluidics which is a group of technologies that exhibit moderate throughput with the capability to address droplets in series. Adapted with permission from [165].

Individual droplets can be handled, for example, using acoustophoresis/SAW [134,167], hydrodynamic or microfluidic changes by geometries [168–171], elective emulsion separation [172], in-line passive filters [173], or even DLD [99]. It is also possible to manipulate the contents inside these droplets using DEP [174,175], magnetic fields [176,177], or beads and microfluidic ratchets [178].

Different from that is digital microfluidics (DMF), which employs the electro-wetting effect. Using an external electric field, the interface energy between the polar (water) droplet and the (dielectric) surface can be locally modulated, and thus the contact angle between the droplet and the surface can be reduced. This effectively renders the surface locally more hydrophilic and lets the droplet migrate along this hydrophilicity gradient. Individual droplets can be moved and directed like this, separated, merged or stored as illustrated in Figure 9b. One commercial DMF platform for diagnostics in newborns and children has just been presented [179].

As a general note: droplet microfluidics can be used in many instances like traditional well-plate assays, with the few following adjustments: Well plates are static and highly parallelizable, while droplet microfluidics are more dynamic (i.e., the well plates rest on their plates, while the droplets generally have to move in sequence to the next operational point) and can be handled in high-throughput series. However, especially in drug assays, (droplet) microfluidics-based assays have a great potential to surpass well-plate-based approaches [63].

For an overview of the pros and cons of droplet microfluidics for single cell separation, analysis and diagnostics, please refer to Table 8 below.

Table 8. Pros and cons of droplet microfluidics-based approaches.

Pros	Cons
High throughput and parallelization possible	Additional hydrophobic phase needed, plus either detergent (surfactant) or electrowetting
Reactions can be triggered and their process studied	Microfluidic devices are more complex and thus far cannot be brought outside a lab
Entire libraries of drug targets can be screened	-
Droplets can be stored in loops for long term studies	-
Can be combined with other techniques	-

2.2.2. Optical Density/Refractive Index

Using the refractive index of a single cell within a microchip for potential single cell diagnostics was reported over a decade ago, e.g., by Liang et al. [180]. Over the past 15 years, several applications have been reported, based on advanced measurements of refractive indices of single cells. The main measure of their precision is the Refractive Index Unit (RIU). The refractive index is the ratio of how fast light travels in vacuum to how fast light travels in a medium (e.g., water, cytoplasm). Proteins and DNA have a higher refractive index than water, and thus the local refractive index of a cell can be used to measure the local concentration of protein at any volume of a cell. The refractive index itself has no unit or dimension. The RIU can be seen as the (smallest) portion of (local) change in the refractive index that can be measured by any given method.

Exemplary methods for measuring the refractive indices of single cells onboard of microfluidic chips include light scattering [140], phase contrast microcopy [181], laser resonant cavity [180], Fabry-Pérot cavity [182], Mach-Zehnder interferometry [183], or combining an optical trap with a grating resonant cavity [184].

To generate a combined separation and analysis on-chip device, any of the above-mentioned separation approaches can be combined with any measurement of the refractive index of the isolated cells. The refractive index of a cell is a key biophysical parameter and correlates to biophysical properties including mechanical, optical and electrical properties [185]. With the combination of microfluidics, photonic and imaging technologies, it is now possible to study the 3D refractive index of a cell in the sub-micron regime, as Liu et al. detailed in their review in 2016 [185]. In addition to this, it is possible to use Brillouin microscopy, where the optical phase shift of a cell gives information about the cells local stiffness and water content, given the local refractive indices are known [186], or measured alongside using one of the techniques listed in Table 1. Three major approaches for probing the cell refractive index can be summarized, as shown in Table 9, below:

Table 9. Techniques to measure the cell refractive index, sorted by approach.

Approach	Measurement Technique	Refractive Index Resolution	Spatial Resolution (nm)	Minimal Mass Density Change (g/mL)[1]	Ref.
Bulk (Average refractive index of suspended cells)	Interference refractometer	3×10^{-3}	NA	0.0163	[187]
	Light scattering	1×10^{-2}	NA	0.0542	[188]
	Light transmission and reflection	1×10^{-2}	NA	0.0542	[187]
	Optical densitometer	3×10^{-3}	NA	0.0163	[189]
Single cell level	Fabry-Pérot resonant cavity	3×10^{-3}	NA	0.0163	[182]
	Grating resonant cavity	1×10^{-3}	NA	0.0054	[184]
	Immersion refractometer	1×10^{-3}	NA	0.0054	[190]
	Laser resonant cavity	4×10^{-3}	NA	0.0217	[180]
	Light scattering	1×10^{-2}	NA	0.0542	[140]
	Mach-Zehnder Interferometer	1×10^{-3}	NA	0.0054	[183]
(Sub-)cellular refractive index mapping	Common-path tomographic diffractive microscopy	1×10^{-3}	NA	0.0054	[191]
	Confocal quantitative phase microscopy	4×10^{-3}	NA	0.0217	[164]
	Digital holographic microscopy	3×10^{-4}	NA	0.0016	[163]
		1×10^{-2}	NA	0.0542	[192]
	Hilbert phase microscopy	2×10^{-3}	1000	0.0108	[193]
	Microfluidic off-axis holography	5×10^{-3}	350	0.0217	[194]
	Phase-shifting interferometry	9×10^{-3}	250	0.0488	[195]
		3×10^{-4}	NA	0.0016	[196]
	Surface Plasmon nano-optical probe	4×10^{-5}	80	0.0002	[197,198]
	Tomographic bright-field imaging	8×10^{-3}	260	0.0434	[199]

[1] Values were calculated by Liu et al. in [185].

For a more in-depth discussion on these techniques, see the review by Liu et al. in [185]. A general overview of the pros and cons of these techniques can be found in Table 10 below.

Table 10. Pros and cons of cell refractive index-based approaches.

Pros	Cons
Very adaptable to target cells in general	Has to be combined with separation techniques
Generally can be used for high-throughout	Additional setups can be expensive, depending on the technique employed
Depending on the technique, the needed machinery might already be present for the setup	-
Generally damage-free to sample cells	-
Can be combined with other techniques	-

2.2.3. Paper Microfluidics

Paper microfluidics (PMF) was first used for portable diagnostics in 2007 [200] as a low-cost alternative to continuous phase or droplet microfluidics. Instead of using walls and hollow structures, paper microfluidics uses paper as a hydrophilic stationary phase (that carries the fluid) and a hydrophobic phase (paper treated with wax, photoresist, graphene or other substances) to block the fluid. Since paper is, on the microscopic scale, a tangle of fibers, most cells and similarly big particles are retained, while the smaller and soluble parts travel along the paper as they do in thin layer chromatography (see Figure 10). Thus, PMF is used more often to analyze the contents of lysed cells (e.g., DNA [11,16,20]) than entire cells. However, it also possible to have a PMF device that starts out with a sample of cells and media, first separates the cells, then lyses the isolated cells and runs molecular analysis on them. One example is a field-applicable test using foldable paper slips for diagnosis of Malaria in infected blood samples [11].

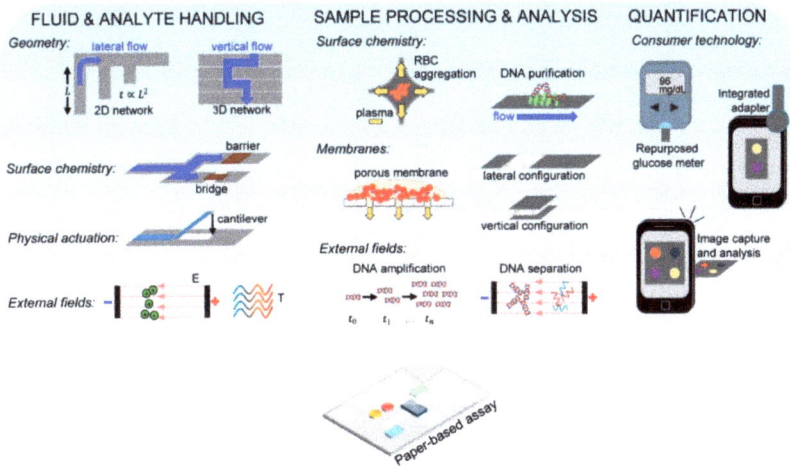

Figure 10. Paper microfluidic applications for diagnostics, including fluid and analyte handling, sample processing and analysis, as well as quantification. Reproduced with permission from [33].

The adaptability and compatibility of PMF with smart phone analysis, the mass-producibility and readily available and cheap materials, along with the possibility to make three-dimensional devices by folding, cutting, and washing the paper allowed paper microfluidics to make enormous advances in the last 12 years. From sample collection to signaling the result, paper-based assays have been used for every step along diagnostic pipeline, generating "a formidable toolbox of advanced strategies for fluid and analyte manipulation in paper-based assays" [33]. With the appropriate (pre-)treatments, paper-based assays allow portable and hand-held versions of analytical techniques, such as DNA separation and nucleic acid amplification, outside of specialized laboratories, even unlocking them for field tests in low- and middle-income countries.

While paper-based microfluidic assays are a cheap and reliable alternative to other single cell level diagnostic devices, they need a different approach than other microfluidic techniques and also face specific challenges especially along their translation and commercialization: (1) Integrating the multiple tasks into one single system and (2) obtaining clinical validation, while (3) adhering to the various EU/US/national regulations, (4) up-scaling of production, especially with multiple pre-treatments on the paper, and (5) maintaining storability. All this notwithstanding, PMF is arguably the leading approach to single cell diagnostic chips, as shown by many reviews over the last two decades detailing the usages of PMF diagnostic chips, also called microfluidic analytic devices (µPADs) [20,33,78,201–205]. PMF has been combined with many other techniques to create a plethora of applications: from combining PMF with immunoassays [206] for Hepatitis C-tests, via on-board batteries [207] or cell phones [203] to generate field-applicable µPADs, to proteomics [208] parasite diagnostics. How established µPADs are in our everyday life is demonstrated by the usage of paper microfluidics for cheap home tests for male (and female) fertility and sperm DNA integrity [16–20].

The main difference in cell handling in PMF compared to continuous phase or droplet microfluidics is that cells generally do not travel through or along the paper. Individual molecules (e.g., after the cell was lysed), however, can diffuse through the paper in a fashion very similar to (thin layer) chromatography. This also opens up targeted retention of cells during washing steps (e.g., [11]).

For an overview of the pros and cons of PMF in the context of single cell separation and diagnostics, please refer to Table 11 below.

Table 11. Pros and cons of paper microfluidics.

Pros	Cons
Can integrate separation and detection	No continuous phase or separation of cells
Can be used for complete diagnostic chips	Not usable for high throughput
Potentially cheap enough for disposable tests	Parallelization limited
Disposable tests can be combined with analytic cassettes to allow mass field testing with combined lab-based analysis	Slower than liquid–liquid microfluidics
Very mature technology with many applications	Mostly single use devices

2.3. Molecular Analysis of Single Cells

If not only the cell as a whole, but also subcellular and molecular parts of the cell, are the target of a diagnostic chip, it is a standard step to lyse the cells, wash, filter, optionally amplify selected parts and finally verify the existence and concentration of the molecules of interest. For this molecular analysis of single cells, the following technologies are most commonly used in today's diagnostic chips:

2.3.1. Polymerase Chain Reaction (PCR, Nested PCR, qPCR/ RT-PCR)

Especially after lysing cells to check for intracellular or subcellular markers (e.g., DNA, RNA fragments, antibodies), the sheer plethora of compounds can be hard to read out when looking for a single structure. In this case, an amplification step is introduced to replicate the specific parts of this wild mixture of substances. The most basic technique, polymerase chain reaction (PCR) simply emulates a process that happens within us every day (see Figure 11). To verify the presence of a suspected target (e.g., the Malaria causing parasite *Plasmodium falciparum*), all the DNA in the cell is denaturized (basically unfolded from the double helix). Since the DNA of individual life forms (e.g., *P. falciparum*) have distinctive already-known sequences, we can select these sequences to be multiplied by adding primers, consisting of the corresponding base pairs (see Figure 11).

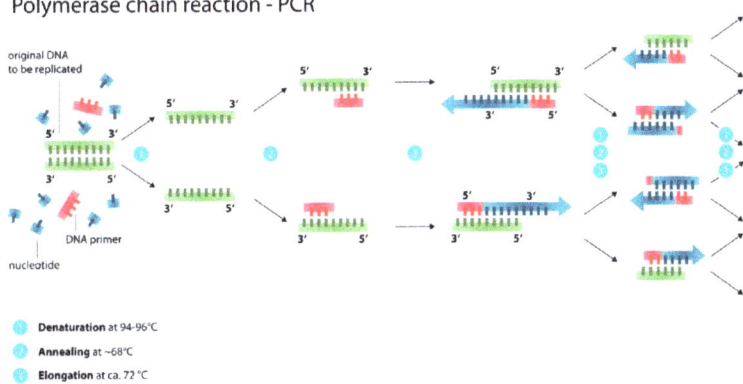

Figure 11. Scheme of the polymerase chain reaction: (**1**) Denaturation: The original DNA double strand is split into two complementary strands. While in our bodies this happens enzymatically at our body temperature, in microfluidic chips the DNA has to be heated to about 95 °C. (**2**) Annealing: Compatible DNA primers attach to the individual DNA single strands. These normally consist of several bases to increase the chances that this primer is specific and attaches as close to the end of the single DNA strands as possible. On chip this is done around 65–68 °C (**3**) Elongation: Beginning from the 3′-end of the DNA primer, individual nucleotides (bases) attach complementary to the respective single stranded DNA. Thereby, Arginine and Tyrosine complement each other and Cytosine and Guanine respectively. Elongation on chip usually takes place at around 72 °C. Images were adapted from Wikipedia user Enzoklop in 2014.

Using an enzymatic cleaving of the double stranded DNA into single strands allows individual nucleotides to assemble on these single strands and form two new double stranded DNA molecules. While our bodies can perform this at body temperature, on a chip, this amplification loop is carried out at higher temperatures: cleaving of double stranded DNA (Denaturation) at 94–96 °C; recombination of single strands with matching primers (Annealing) at ca. 68 °C; and the adding of further bases (Elongation) at ca. 72 °C. Going through this amplification loop (see Figure 11 (1–3)) once doubles the amount of targeted DNA. However, with each copy, the strand becomes shorter by a basis. This happens both in vitro and in our bodies and is believed to be a main factor of aging [209].

Within the confines of a microfluidic chip, such different temperatures can best be achieved by having a long channel meandering over different heating stages, as demonstrated exemplarily by Ma et al. in 2019 [210]. This can also be done with droplets that already contain the primers and nucleotides [210]. The application of PCR is especially widely applied for diagnostic chips for malaria detection [11,13,211] and other parasites [43,212].

With our growing understanding of the biophysics and biochemistry of life, the PCR technique has been advanced. Nowadays, it is commonplace to already obtain real-time quantified results (real-time PCR; RT-PCR or quantitative PCR, qPCR) in between the amplification steps. Basically, this is done by adding a fluorophore to the DNA-primers (or other prominent parts), which emits a fluorescent signal only when bound to a strand of DNA (or which stops emitting after being bound to other nucleotides). By comparing the intensity of fluorescent signal after each amplification loop, it is possible to quantify the amount of target DNA within the sample. The incorporation of qPCR into single cell diagnostic chips has recently been discussed more in detail by Reece et al. [37]. In summary, nowadays, microfluidic single cell separations based on optical manipulation, microfluidic large-scale integration, hydrodynamic cell sorting/stretching, and droplet microfluidics, have achieved the high-throughput capacities necessary to allow—in combination with PCR-based analysis—population-wide screenings of samples on the single cellular level.

To reduce non-specific binding in PCR products that arise from unintended primer binding sites, one loop of PCR can be nested inside another, using a different set of primers. This nested-PCR is also a widespread practice in (single cell) diagnostic chips and its application has recently been reviewed and discussed with other PCR techniques [213].

A combined overview of the pros and cons of both PCR and LAMP-based techniques can be found in Table 12 below.

Table 12. Pros and cons of PCR/LAMP-based analysis.

Pros	Cons
Reliable	Needs electricity/highly controlled heat source
Can be combined for diagnostic chips	Needs computer hard and software for quantitative tests
Parallelizable and ready for high throughput	Sensitive to impurities and contamination
To some extent, test kits can be pre-treated with reactants to enable field-applicable testing	-
Can be combined with all above separation techniques, including paper microfluidics	-

2.3.2. Loop-Mediated Isothermal Amplification (LAMP, NAAT, LAMPport, NINA-LAMP)

An alternative technique for diagnosing sub-cellular targets is loop-mediated isothermal amplification (LAMP), which can run at a single temperature of around 65 °C and inside a single tube where the cell lysate is mixed with DNA polymerase and a set of four (or more) specifically designed primers [214]. The fact that it needs less equipment to be run, and that it can even be more portable [55], or fully non-instrumented [48], makes LAMP and its derivatives very promising for use in the field for resource-scarce settings [11,22,47,48,55,61].

One example for a field-applicable LAMP device is a chip for nucleic acid amplification test (NAAT, see Figure 12). In brief, operating the NAAT consists of the following steps: First, the cells are lysed and the lysate is introduced into the chip. Second, a buffer is used to wash the lysate and stabilize the targeted molecules. Third, water is added to the solution, which may already contain reagents and enzymes. Fourth, by heating the reaction volume to 65 °C, the amplification loop is started. In some cases, the heater also melts away an encapsulation between the (lyophilized) enzymes and reagents and the water [61]. Finally, after the amplification has run for the desired amount of time, the reaction mixture is excited with a light source and the fluorescent response of the reaction mixture is measured to assess the presence and concentration of the target [61].

Even more advanced and exciting are LAMP-derived techniques, specifically designed for the deployment in areas with very limited access to resources, e.g., field work in endemic areas of parasitic diseases like Malaria, LAMPport (portable) [55] and non-instrumented nucleic acid amplification LAMP (NINA-LAMP) [48]. With further tweaking, it is possible to attain field-applicable multi-parasite or multi-disease tests at low costs—if it can be combined with a proper separation technique, like DLD or paper microfluidics.

2.3.3. Proteomics, Metabolomics, Transcriptomics and Polyomics

A very important set of techniques for diagnostic chips is the group of prote-/metabol-/transcript-/gen- or poly 'omics'. According to Haring and Wallaschofksi, "omic-metrics including the Phenome (physical traits such as body height, weight, or specific personality characteristics), Metabolome (complete set of small-molecule metabolites to be found within a bio- logical sample), Proteome (entire set of proteins expressed by a genome, cell, tissue, or organism), Transcriptome (information about the expression of individual genes at the messenger ribonucleic acid level), Genome (complete set of genes in the [...] organism)" [215]. Single cell metabolomics is considered a crucial element for targeted drug discovery [216], and already ten years ago, single cell analysis was the new frontier in 'omics' for Wang and Bodovitz [217].

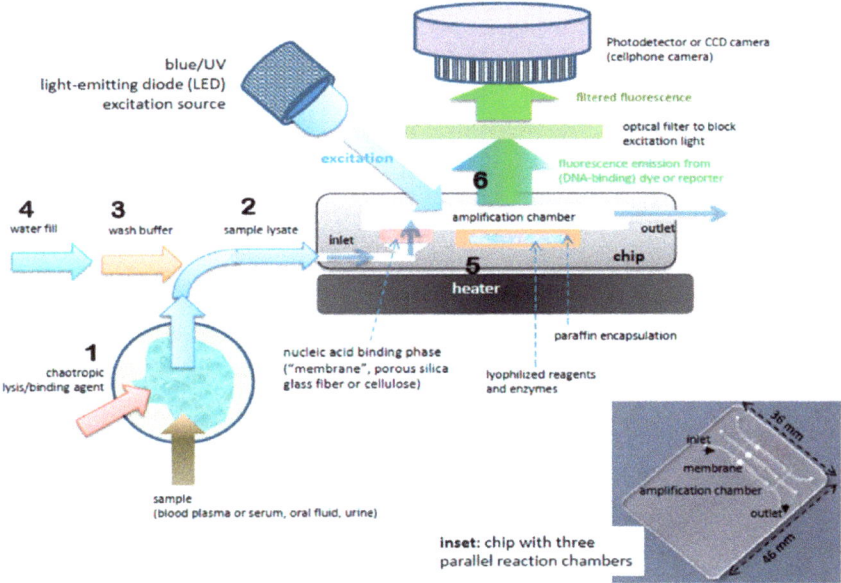

Figure 12. Process steps and (cross-section) schematic of chip for nucleic acid amplification test (NAAT). Chip has one or more flow-through chambers for isothermal amplification and includes a filter-like, flow-porous nucleic acid binding phase (e.g., silica glass fiber or cellulose) and is pre-loaded with paraffin-encapsulated amplification reagents (lyophilized polymerase, primers, fluorescence reporter DNA-intercalating, dyes, and other components). Operational steps: (**1**) sample is mixed off-chip with lysis/binding reagent buffer (containing e.g., chaotropic agent such as guanidinium HCl) that lyses virus and cells and promotes nucleic acid adsorption to binding media, e.g., silica glass fiber or cellulose ('membrane'), (**2**) sample (~100 µL) is injected into chip with pipette or syringe, (**3**) ethanol-based, high-salt buffer (~100 µL) is injected into chip to wash the membrane (keeping most of the captured nucleic acid adsorbed to the membrane, (**4**) chamber (25 to 50 µL volume) is filled with water and sealed with tape, (**5**) chip is heated to amplification temperature (~65 °C) using a small (~1 Watt) electric-heater. The heating melts the paraffin encapsulation, releasing and reconstituting the reagents, (**6**) the amplification reaction is excited with a blue or UV LED, such that the DNA intercalating dye generates a fluorescence signal proportional to the amount of DNA amplicon produced. The fluorescence is measured by filtering the excitation light and detection with a photodetector of CCD camera, such as provided by a mounted cellphone. Reproduced with permission from [61].

In general, all single cellular 'omics' share a basic setup: a target cell (or several) is lysed, their contents may be treated and are chromatographically separated and analyzed using mass-spectroscopy (or similar techniques). Most often, separation and detection is done using gas chromatography with mass spectroscopy (GC-MS) [218], high-pressure liquid chromatography and mass spectroscopy (HPLC-MS) [219], and capillary electrophoresis with mass spectroscopy (CE-MS) [220]. Since a single cell contains a plethora of substances that can be found after the chromatography in the spectra, it is usually not possible to identify individual peaks, but rather "fingerprints" of overlying bands. By specifically adding individual molecules to this cellular cocktail, changes in the overlying bands can be assigned individual substances in a given setting. With careful sample treatment, -omics can be very well used for diagnostic chips and to unravel to exact roles of individual substances within these cells, their metabolism, their transcription during cell division, or all of the above. Proteomics has been massively employed to elucidate the host interactions of SARS-CoV-2, the causative agent of COVID-19 [27–32], the greatest pandemic in our millennium.

For an overview of the pros and cons of employing polyomics towards single cell level diagnostics, please refer to Table 13 below.

Table 13. Pros and cons of omics-based analysis.

Pros	Cons
Can give a complete picture of intracellular life	Needs electricity
Can be combined for diagnostic chips	Needs computer hard and software for qualitative and quantitative tests
Generally ready for high throughput	Sensitive to impurities and contamination
Can be used to find out changes (due to stimuli) in complex samples like tissues	Needs highly trained personnel
Can be combined with all above separation techniques, limited with paper microfluidics	Most commercial setups lack the flexibility to adapt the setup (i.e., change eluents) to improve separation of non-standard target molecules

3. Implementation

3.1. Point-of-Care Diagnostics (POC)

Point-of-Care (POC) means that these diagnostic devices can be used at the patient level and deliver a diagnosis locally and quickly—without need for taking samples, sending them away for analysis and having to continue care while waiting on the outcome of the diagnostic test. This is often achieved with the LAMP test and its derivatives, LAMPport and NINA-LAMP [22,47,48,55,61]. However, many other, especially microfluidic, approaches have also been tested and discussed for their aptitude to be used for POC diagnostic tools [38,45,205,212,221–225].

3.2. Biosensors in the Developing World and the Need for ASSURED

Many publications, which demonstrate microfluidic test devices mention at least some of the WHO's "ASSURED" criteria (i.e., Affordable, Sensitive, Specific, User-friendly, Rapid and robust, Equipment-free, Delivered to the users who need them) for low-cost sensors for the developing world. While the functionality "ASSR" of the low-cost sensors is often easily achieved, their ready user acceptance and field-applicability "UED" often lags behind [201].

Microfluidics devices make it possible to carry out complex analysis outside of highly equipped laboratories, often in a hand-held, portable, and field-applicable fashion [226], at least that is the promise of the "lab on a chip" idea. In reality, for many steps along the way, we experts experience it to be a "chip in a lab" first, with many hurdles to overcome, including the "valley of death": the gap between the academic research and industrial/real-world application.

4. Discussion

Given the steady increase in publications and research carried out towards the creation and usage of single cell diagnostic chips, it can really be regarded as a hot topic within the microfluidics community. On the diagnostic side, single cell diagnostics means that infections and diseases can already be diagnosed with a minimal sample volume and from a single pathogenic cell. On the production side, this means that high-resolution techniques have to be used (e.g., soft lithography), but also that it is possible to parallelize and mass produce these microfluidic chips.

The vast adaptability of microfluidics to other techniques and technologies to handle and characterize cells, from bulk to individual, opens new possibilities and synergies every year. While this plethora of opportunities continues to be explored, it is always a rather long stretch from the "chip in a lab" to the field-applicable "lab on a chip". Many great diagnostic chips that could help study, diagnose and eventually eradicate diseases—especially neglected tropical diseases (NTDs), do not make it to the final deployment in the field. It is a sad state of affairs and a symptom of some mismatches in the research and funding landscape, that many projects aiming to understand the fundamentals are funded, and even more projects are funded to develop actual applications and physical single cell diagnostic chips, but that only very few devices ever make it across this "valley of death" to their field application. Several funding bodies like the Bill and Melinda Gates Foundation or the UK's Grand Challenges Research Fund support research to develop such chips for low- and middle-income countries. However, the—now—most important step to get the chips mass produced, certified and deployed, is sadly virtually unfunded; an immense potential to actually make a positive impact on the world is not being seized.

5. Conclusions

In conclusion, there have been many technologies that can be combined to create powerful single cell diagnostic chips, but without a massive change in the system, many of these chips will remain "chips in a lab" rather than unleashing their true potential as "labs on a chip" at the point of care.

Funding: The APC was funded by the German Federal Ministry of Education and Research Project Nanomatfutur FKZ13N13547.

Acknowledgments: The author wishes to thank Elke Neu-Ruffing for her exemplary support and guidance.

Conflicts of Interest: The author declares no conflict of interest.

References

1. Nematbakhsh, Y.; Lim, C.T. Cell biomechanics and its applications in human disease diagnosis. *Acta Mech. Sin.* **2015**, *31*, 268–273. [CrossRef]
2. Kozminsky, M.; Wang, Y.; Nagrath, S. The incorporation of microfluidics into circulating tumor cell isolation for clinical applications. *Curr. Opin. Chem. Eng.* **2016**, *11*, 59–66. [CrossRef]
3. Okano, H.; Konishi, T.; Suzuki, T.; Suzuki, T.; Ariyasu, S.; Aoki, S.; Abe, R.; Hayase, M. Enrichment of circulating tumor cells in tumor-bearing mouse blood by a deterministic lateral displacement microfluidic device. *Biomed. Microdevices* **2015**, *17*, 1–11. [CrossRef]
4. Au, S.H.; Edd, J.; Stoddard, A.E.; Wong, K.H.K.; Fachin, F.; Maheswaran, S.; Haber, D.A.; Stott, S.L.; Kapur, R.; Toner, M. Microfluidic isolation of circulating tumor cell clusters by size and asymmetry. *Sci. Rep.* **2017**, *7*. [CrossRef]
5. Liu, Z.; Zhang, W.; Huang, F.; Feng, H.; Shu, W.; Xu, X.; Chen, Y. High throughput capture of circulating tumor cells using an integrated microfluidic system. *Biosens. Bioelectron.* **2013**, *47*, 113–119. [CrossRef]
6. Aghaamoo, M.; Aghilinejad, A.; Chen, X.; Xu, J. On the design of deterministic dielectrophoresis for continuous separation of circulating tumor cells from peripheral blood cells. *Electrophoresis* **2019**, *40*, 1486–1493. [CrossRef]
7. Kong, T.F.; Ye, W.; Peng, W.K.; Hou, H.W.; Preiser, P.R.; Nguyen, N.-T.; Han, J. Enhancing malaria diagnosis through microfluidic cell enrichment and magnetic resonance relaxometry detection. *Sci. Rep.* **2015**, *5*, 11425. [CrossRef] [PubMed]
8. Barrett, M.P.; Cooper, J.M.; Regnault, C.; Holm, S.H.; Beech, J.P.; Tegenfeldt, J.O.; Hochstetter, A. Microfluidics-Based Approaches to the Isolation of African Trypanosomes. *Pathogens* **2017**, *6*, 47. [CrossRef] [PubMed]
9. Holm, S.H. *Microfluidic Cell and Particle Sorting Using Deterministic Lateral Displacement*; Lund University: Lund, Sweden, 2018.
10. Clark, D.J.; Moore, C.M.; Flanagan, M.; Van Bocxlaer, K.; Piperaki, E.-T.; Yardley, V.; Croft, S.L.; Tyson, J.; Whitehouse, S.P.; O'Halloran, J.; et al. An efficient and novel technology for the extraction of parasite genomic DNA from whole blood or culture. *Biotechniques* **2019**, *68*, 1–6. [CrossRef]
11. Reboud, J.; Xu, G.; Garrett, A.; Adriko, M.; Yang, Z.; Tukahebwa, E.M.; Rowell, C.; Cooper, J.M. Paper-based microfluidics for DNA diagnostics of malaria in low resource underserved rural communities. *Proc. Natl. Acad. Sci. USA* **2019**, *116*, 4834–4842. [CrossRef] [PubMed]

12. McGrath, J.S.; Honrado, C.; Spencer, D.; Horton, B.; Bridle, H.L.; Morgan, H. Analysis of Parasitic Protozoa at the Single-cell Level using Microfluidic Impedance Cytometry. *Sci. Rep.* **2017**, *7*, 1–11. [CrossRef] [PubMed]
13. Nam, J.; Shin, Y.; Tan, J.K.S.; Lim, Y.B.; Lim, C.T.; Kim, S. High-throughput malaria parasite separation using a viscoelastic fluid for ultrasensitive PCR detection. *Lab Chip* **2016**, *16*, 2086–2092. [CrossRef] [PubMed]
14. Warkiani, M.E.; Tay, A.K.P.; Khoo, B.L.; Xiaofeng, X.; Han, J.; Lim, C.T. Malaria detection using inertial microfluidics. *Lab Chip* **2015**, *15*, 1101–1109. [CrossRef] [PubMed]
15. Kolluri, N.; Klapperich, C.M.; Cabodi, M. Towards lab-on-a-chip diagnostics for malaria elimination. *Lab Chip* **2017**, *18*, 75–94. [CrossRef]
16. Gong, M.M.; Nosrati, R.; San Gabriel, M.C.; Zini, A.; Sinton, D. Direct DNA Analysis with Paper-Based Ion Concentration Polarization. *J. Am. Chem. Soc.* **2015**, *137*, 13913–13919. [CrossRef] [PubMed]
17. Nosrati, R.; Gong, M.M.; Gabriel, M.C.S.; Pedraza, C.E.; Zini, A.; Sinton, D. Paper-based quantification of male fertility potential. *Clin. Chem.* **2016**, *62*, 458–465. [CrossRef]
18. Coppola, M.A.; Klotz, K.L.; Kim, K.A.; Cho, H.Y.; Kang, J.; Shetty, J.; Howards, S.S.; Flickinger, C.J.; Herr, J.C. SpermCheck® Fertility, an immunodiagnostic home test that detects normozoospermia and severe oligozoospermia. *Hum. Reprod.* **2010**, *25*, 853–861. [CrossRef]
19. Matsuura, K.; Chen, K.H.; Tsai, C.H.; Li, W.; Asano, Y.; Naruse, K.; Cheng, C.M. Paper-based diagnostic devices for evaluating the quality of human sperm. *Microfluid. Nanofluid.* **2014**, *16*, 857–867. [CrossRef]
20. Nosrati, R.; Gong, M.M.; San Gabriel, M.C.; Zini, A.; Sinton, D. Paper-based sperm DNA integrity analysis. *Anal. Methods* **2016**, *8*, 6260–6264. [CrossRef]
21. Birch, C.M.; Hou, H.W.; Han, J.; Niles, J.C. Identification of malaria parasite-infected red blood cell surface aptamers by inertial microfluidic SELEX (I-SELEX). *Sci. Rep.* **2015**, *5*, 1–16. [CrossRef]
22. Song, J.; Liu, C.; Bais, S.; Mauk, M.G.; Bau, H.H.; Greenberg, R.M. Molecular Detection of Schistosome Infections with a Disposable Microfluidic Cassette. *PLoS Negl. Trop. Dis.* **2015**, *9*, 1–18. [CrossRef] [PubMed]
23. Pham, N.M.; Rusch, S.; Temiz, Y.; Beck, H.P.; Karlen, W.; Delamarche, E. Immuno-gold silver staining assays on capillary-driven microfluidics for the detection of malaria antigens. *Biomed. Microdevices* **2019**, *21*. [CrossRef] [PubMed]
24. Beech, J.P.; Ho, B.D.; Garriss, G.; Oliveira, V.; Henriques-Normark, B.; Tegenfeldt, J.O. Separation of pathogenic bacteria by chain length. *Anal. Chim. Acta* **2018**, *1000*, 223–231. [CrossRef]
25. Tottori, N.; Nisisako, T.; Park, J.; Yanagida, Y.; Hatsuzawa, T. Separation of viable and nonviable mammalian cells using a deterministic lateral displacement microfluidic device. *Biomicrofluidics* **2016**, *10*, 014125. [CrossRef]
26. Beeman, A.Q.; Njus, Z.L.; Pandey, S.; Tylka, G.L. Chip technologies for screening chemical and biological agents against plant-parasitic nematodes. *Phytopathology* **2016**, *106*, 1563–1571. [CrossRef] [PubMed]
27. Wang, H.; Hou, X.; Wu, X.; Liang, T.; Zhang, X.; Wang, D.; Teng, F.; Dai, J.; Duan, H.; Guo, S.; et al. SARS-CoV-2 proteome microarray for mapping COVID-19 antibody interactions at amino acid resolution. *bioRxiv* **2020**. [CrossRef]
28. Jenkins, C.; Orsburn, B. In silico approach to accelerate the development of mass spectrometry-based proteomics methods for detection of viral proteins: Application to COVID-19. *bioRxiv* **2020**. [CrossRef]
29. Al-Motawa, M.; Abbas, H.; Wijten, P.; de la Fuente, A.; Xue, M.; Rabbani, N.; Thornalley, P.J. Vulnerabilities of the SARS-CoV-2 virus to proteotoxicity—Opportunity for repurposed chemotherapy of COVID-19 infection. *bioRxiv* **2020**. [CrossRef]
30. Giri, R.; Bhardwaj, T.; Shegane, M.; Gehi, B.R.; Kumar, P.; Gadhave, K. Dark proteome of Newly Emerged SARS-CoV-2 in Comparison with Human and Bat Coronaviruses. *bioRxiv* **2020**. [CrossRef]
31. Liang, Q.; Li, J.; Guo, M.; Tian, X.; Liu, C.; Wang, X.; Yang, X.; Wu, P.; Xiao, Z.; Qu, Y.; et al. Virus-host interactome and proteomic survey of PMBCs from COVID-19 patients reveal potential virulence factors influencing SARS-CoV-2 pathogenesis. *bioRxiv* **2020**. [CrossRef]
32. Gordon, D.E.; Jang, G.M.; Bouhaddou, M.; Xu, J.; Obernier, K.; O'Meara, M.J.; Guo, J.Z.; Swaney, D.L.; Tummino, T.A.; Hüttenhain, R.; et al. A SARS-CoV-2-Human Protein-Protein Interaction Map Reveals Drug Targets and Potential Drug-Repurposing. *bioRxiv* **2020**. [CrossRef]
33. Gong, M.M.; Sinton, D. Turning the Page: Advancing Paper-Based Microfluidics for Broad Diagnostic Application. *Chem. Rev.* **2017**, *117*, 8447–8480. [CrossRef]

34. Roberts, S.F.; Fischhoff, M.A.; Sakowski, S.A.; Feldman, E.L. Perspective: Transforming science into medicine: How clinician-scientists can build bridges across research's "valley of Death". *Acad. Med.* **2012**, *87*, 266–270. [CrossRef]
35. Bottazzi, M.; Dumonteil, E.; Valenzuela, J.; Betancourt-Cravioto, M.; Tapia-Conyer, R.; Hotez, P. Bridging the innovation gap for neglected tropical diseases in México: Capacity building for the development of a new generation of antipoverty vaccines. *Bol. Med. Hosp. Infant. Mex.* **2011**, *68*, 138–146.
36. Sarkar, S.; Sabhachandani, P.; Konry, T. Isothermal Amplification Strategies for Detection in Microfluidic Devices. *Trends Biotechnol.* **2017**, *35*, 186–189. [CrossRef]
37. Reece, A.; Xia, B.; Jiang, Z.; Noren, B.; McBride, R.; Oakey, J. Microfluidic techniques for high throughput single cell analysis. *Curr. Opin. Biotechnol.* **2016**, *40*, 90–96. [CrossRef]
38. Esfandyarpour, R.; DiDonato, M.J.; Yang, Y.; Durmus, N.G.; Harris, J.S.; Davis, R.W. Multifunctional, inexpensive, and reusable nanoparticle-printed biochip for cell manipulation and diagnosis. *Proc. Natl. Acad. Sci. USA* **2017**, *114*, E1306–E1315. [CrossRef]
39. Zhang, S.; Tis, T.B.; Wei, Q. *Smartphone-Based Clinical Diagnostics. Precision Medicine for Investigators, Practitioners and Providers*; Elsevier: Amsterdam, The Netherlands, 2020.
40. Civin, C.I.; Ward, T.; Skelley, A.M.; Gandhi, K.; Peilun Lee, Z.; Dosier, C.R.; D'Silva, J.L.; Chen, Y.; Kim, M.J.; Moynihan, J.; et al. Automated leukocyte processing by microfluidic deterministic lateral displacement. *Cytom. Part A* **2016**, *89*, 1073–1083. [CrossRef]
41. Pollak, J.J.; Houri-Yafin, A.; Salpeter, S.J. Computer Vision Malaria Diagnostic Systems—Progress and Prospects. *Front. Public Health* **2017**, *5*, 1–5. [CrossRef]
42. Lee, W.G.; Kim, Y.G.; Chung, B.G.; Demirci, U.; Khademhosseini, A. Nano/Microfluidics for diagnosis of infectious diseases in developing countries. *Adv. Drug Deliv. Rev.* **2010**, *62*, 449–457. [CrossRef]
43. Phuakrod, A.; Sripumkhai, W.; Jeamsaksiri, W.; Pattamang, P.; Juntasaro, E.; Thienthong, T.; Foongladda, S.; Brindley, P.J.; Wongkamchai, S. Diagnosis of feline filariasis assisted by a novel semi-automated microfluidic device in combination with high resolution melting real-time PCR. *Parasites Vectors* **2019**, *12*, 1–9. [CrossRef]
44. Saeed, M.A.; Jabbar, A. "Smart Diagnosis" of Parasitic Diseases by Use of Smartphones. *J. Clin. Microbiol.* **2017**, *56*, e01469-17. [CrossRef]
45. Ruiz-Vega, G.; Arias-Alpízar, K.; De Serna, E.; Borgheti-Cardoso, L.N.; Sulleiro, E.; Molina, I.; Fernàndez-Busquets, X.; Sánchez, A.; Campo, F.J.; Baldrich, E. Electrochemical POC device for fast malaria quantitative diagnosis in whole blood by using magnetic beads, poly-HRP and microfluidic paper electrodes. *Biosens. Bioelectron.* **2019**, *150*, 111925. [CrossRef]
46. Choi, J.; Cho, S.J.; Kim, Y.T.; Shin, H. Development of a film-based immunochromatographic microfluidic device for malaria diagnosis. *Biomed. Microdevices* **2019**, *21*, 86. [CrossRef]
47. Mao, R.; Ge, G.; Wang, Z.; Hao, R.; Zhang, G.; Yang, Z.; Lin, B.; Ma, Y.; Liu, H.; Du, Y. A multiplex microfluidic loop-mediated isothermal amplification array for detection of malaria-related parasites and vectors. *Acta Trop.* **2018**, *178*, 86–92. [CrossRef]
48. Sema, M.; Alemu, A.; Bayih, A.G.; Getie, S.; Getnet, G.; Guelig, D.; Burton, R.; LaBarre, P.; Pillai, D.R. Evaluation of non-instrumented nucleic acid amplification by loop-mediated isothermal amplification (NINA-LAMP) for the diagnosis of malaria in Northwest Ethiopia. *Malar. J.* **2015**, *14*, 1–9. [CrossRef]
49. Juul, S.; Nielsen, C.J.F.; Labouriau, R.; Roy, A.; Tesauro, C.; Jensen, P.W.; Harmsen, C.; Kristoffersen, E.L.; Chiu, Y.L.; Frohlich, R.; et al. Droplet microfluidics platform for highly sensitive and quantitative detection of malaria-causing plasmodium parasites based on enzyme activity measurement. *ACS Nano* **2012**, *6*, 10676–10683. [CrossRef]
50. Fraser, L.A.; Kinghorn, A.B.; Dirkzwager, R.M.; Liang, S.; Cheung, Y.W.; Lim, B.; Shiu, S.C.C.; Tang, M.S.L.; Andrew, D.; Manitta, J.; et al. A portable microfluidic Aptamer-Tethered Enzyme Capture (APTEC) biosensor for malaria diagnosis. *Biosens. Bioelectron.* **2018**, *100*, 591–596. [CrossRef]
51. Hochstetter, A. *Motility, Manipulation and Controlling of Unicellular Organisms*; Universität Basel: Basel, Switzerland, 2014.
52. Hochstetter, A.; Pfohl, T. Motility, Force Generation, and Energy Consumption of Unicellular Parasites. *Trends Parasitol.* **2016**, *32*, 531–541. [CrossRef]
53. Stellamanns, E.; Uppaluri, S.; Hochstetter, A.; Heddergott, N.; Engstler, M.; Pfohl, T. Optical trapping reveals propulsion forces, power generation and motility efficiency of the unicellular parasites Trypanosoma brucei brucei. *Sci. Rep.* **2015**, *4*, 6515. [CrossRef]

54. Menachery, A.; Kremer, C.; Wong, P.E.; Carlsson, A.; Neale, S.L.; Barrett, M.P.; Cooper, J.M. Counterflow dielectrophoresis for trypanosome enrichment and detection in blood. *Sci. Rep.* **2012**, *2*, 1–5. [CrossRef]
55. Wan, L.; Gao, J.; Chen, T.; Dong, C.; Li, H.; Wen, Y.Z.; Lun, Z.R.; Jia, Y.; Mak, P.I.; Martins, R.P. LampPort: A handheld digital microfluidic device for loop-mediated isothermal amplification (LAMP). *Biomed. Microdevices* **2019**, *21*, 9. [CrossRef]
56. Hamula, C.L.A.; Zhang, H.; Li, F.; Wang, Z.; Chris Le, X.; Li, X.-F. Selection and analytical applications of aptamers binding microbial pathogens. *TrAC Trends Anal. Chem.* **2011**, *30*, 1587–1597. [CrossRef]
57. Bourquin, Y.; Syed, A.; Reboud, J.; Ranford-Cartwright, L.C.; Barrett, M.P.; Cooper, J.M. Rare-cell enrichment by a rapid, label-free, ultrasonic isopycnic technique for medical diagnostics. *Angew. Chem. Int. Ed.* **2014**, *53*, 5587–5590. [CrossRef]
58. Nasseri, B.; Soleimani, N.; Rabiee, N.; Kalbasi, A.; Karimi, M.; Hamblin, M.R. Point-of-care microfluidic devices for pathogen detection. *Biosens. Bioelectron.* **2018**, *117*, 112–128. [CrossRef]
59. Shembekar, N.; Chaipan, C.; Utharala, R.; Merten, C.A. Droplet-based microfluidics in drug discovery, transcriptomics and high-throughput molecular genetics. *Lab Chip* **2016**, *16*, 1314–1331. [CrossRef]
60. Zhang, H.; Yang, Y.; Li, X.; Shi, Y.; Hu, B.; An, Y.; Zhu, Z.; Hong, G.; Yang, C.J. Frequency-enhanced transferrin receptor antibody-labelled microfluidic chip (FETAL-Chip) enables efficient enrichment of circulating nucleated red blood cells for non-invasive prenatal diagnosis. *Lab Chip* **2018**, *18*, 2749–2756. [CrossRef]
61. Mauk, M.G.; Song, J.; Liu, C.; Bau, H.H. Simple approaches to minimally-instrumented, microfluidic-based point-of-care Nucleic Acid Amplification Tests. *Biosensors* **2018**, *8*, 17. [CrossRef]
62. Pham, N.M.; Karlen, W.; Beck, H.P.; Delamarche, E. Malaria and the "last" parasite: How can technology help? *Malar. J.* **2018**, *17*, 1–16. [CrossRef]
63. Regnault, C.; Dheeman, D.; Hochstetter, A. Microfluidic Devices for Drug Assays. *High-Throughput* **2018**, *7*, 18. [CrossRef]
64. Hochstetter, A. Presegmentation Procedure Generates Smooth-Sided Microfluidic Devices: Unlocking Multiangle Imaging for Everyone? *ACS Omega* **2019**, *4*, 20972–20977. [CrossRef] [PubMed]
65. Nordenfelt, P.; Cooper, J.M.; Hochstetter, A. Matrix-masking to balance nonuniform illumination in microscopy. *Opt. Express* **2018**, *26*, 17279–17288. [CrossRef] [PubMed]
66. Hochstetter, A.; Stellamanns, E.; Deshpande, S.; Uppaluri, S.; Engstler, M.; Pfohl, T. Microfluidics-based single cell analysis reveals drug-dependent motility changes in trypanosomes. *Lab Chip* **2015**, *15*, 1961–1968. [CrossRef] [PubMed]
67. Wolff, A.; Antfolk, M.; Brodin, B.; Tenje, M. In Vitro Blood-Brain Barrier Models—An Overview of Established Models and New Microfluidic Approaches. *J. Pharm. Sci.* **2015**, *104*, 2727–2746. [CrossRef] [PubMed]
68. Bang, S.; Lee, S.-R.; Ko, J.; Son, K.; Tahk, D.; Ahn, J.; Im, C.; Jeon, N.L. A Low Permeability Microfluidic Blood-Brain Barrier Platform with Direct Contact between Perfusable Vascular Network and Astrocytes. *Sci. Rep.* **2017**, *7*, 8083. [CrossRef] [PubMed]
69. Abhyankar, V.V.; Wu, M.; Koh, C.-Y.; Hatch, A.V. A Reversibly Sealed, Easy Access, Modular (SEAM) Microfluidic Architecture to Establish In Vitro Tissue Interfaces. *PLoS ONE* **2016**, *11*, e0156341. [CrossRef]
70. Van Midwoud, P.M.; Janse, A.; Merema, M.T.; Groothuis, G.M.; Verpoorte, E. Comparison of biocompatibility and adsorption properties of different plastics for advanced microfluidic cell and tissue culture models. *Anal. Chem.* **2012**, *84*, 3938–3944. [CrossRef]
71. Herland, A.; Van Der Meer, A.D.; FitzGerald, E.A.; Park, T.E.; Sleeboom, J.J.F.; Ingber, D.E. Distinct contributions of astrocytes and pericytes to neuroinflammation identified in a 3D human blood-brain barrier on a chip. *PLoS ONE* **2016**, *11*, e0150360. [CrossRef]
72. Helms, H.C.; Abbott, N.J.; Burek, M.; Cecchelli, R.; Couraud, P.O.; Deli, M.A.; Förster, C.; Galla, H.J.; Romero, I.A.; Shusta, E.V.; et al. In vitro models of the blood-brain barrier: An overview of commonly used brain endothelial cell culture models and guidelines for their use. *J. Cereb. Blood Flow Metab.* **2015**, *36*, 862–890. [CrossRef]
73. Adriani, G.; Ma, D.; Pavesi, A.; Goh, E.L.K.; Kamm, R.D. Modeling the Blood-Brain Barrier in a 3D Triple Co-Culture Microfluidic System. In Proceedings of the Annual International Conference of the IEEE Engineering in Medicine and Biology Society, Milan, Italy, 25–29 August 2015; pp. 338–341.
74. Wang, Y.I.; Abaci, H.E.; Shuler, M.L. Microfluidic blood-brain barrier model provides in vivo-like barrier properties for drug permeability screening. *Biotechnol. Bioeng.* **2017**, *114*, 184–194. [CrossRef]

75. Adriani, G.; Ma, D.; Pavesi, A.; Kamm, R.D.; Goh, E.L.K. A 3D neurovascular microfluidic model consisting of neurons, astrocytes and cerebral endothelial cells as a blood-brain barrier. *Lab Chip* **2017**, *17*, 448–459. [CrossRef] [PubMed]
76. Hsieh, C.-C.; Huang, S.-B.; Wu, P.-C.; Shieh, D.-B.; Lee, G.-B. A microfluidic cell culture platform for real-time cellular imaging. *Biomed. Microdevices* **2009**, *11*, 903–913. [CrossRef] [PubMed]
77. Hwang, H.; Park, J.; Shin, C.; Do, Y.; Cho, Y.-K. Three dimensional multicellular co-cultures and anti-cancer drug assays in rapid prototyped multilevel microfluidic devices. *Biomed. Microdevices* **2013**, *15*, 627–634. [CrossRef]
78. Wu, Y.; Gao, Q.; Nie, J.; Fu, J.; He, Y. From Microfluidic Paper-Based Analytical Devices to Paper-Based Biofluidics with Integrated Continuous Perfusion. *ACS Biomater. Sci. Eng.* **2017**, *3*, 601–607. [CrossRef]
79. Wu, J.; Chen, Q.; Liu, W.; He, Z.; Lin, J.M. Recent advances in microfluidic 3D cellular scaffolds for drug assays. *TrAC Trends Anal. Chem.* **2017**, *87*, 19–31. [CrossRef]
80. Hung, P.J.; Lee, P.J.; Sabounchi, P.; Lin, R.; Lee, L.P. Continuous perfusion microfluidic cell culture array for high-throughput cell-based assays. *Biotechnol. Bioeng.* **2005**, *89*, 1–8. [CrossRef]
81. Zhang, B.; Green, J.V.; Murthy, S.K.; Radisic, M. Label-free enrichment of functional cardiomyocytes using microfluidic deterministic lateral flow displacement. *PLoS ONE* **2012**, *7*. [CrossRef]
82. Shao, J.; Wu, L.; Wu, J.; Zheng, Y.; Zhao, H.; Lou, X.; Jin, Q.; Zhao, J. A microfluidic chip for permeability assays of endothelial monolayer. *Biomed. Microdevices* **2010**, *12*, 81–88. [CrossRef]
83. Xu, Z.; Gao, Y.; Hao, Y.; Li, E.; Wang, Y.; Zhang, J.; Wang, W.; Gao, Z.; Wang, Q. Application of a microfluidic chip-based 3D co-culture to test drug sensitivity for individualized treatment of lung cancer. *Biomaterials* **2013**, *34*, 4109–4117. [CrossRef]
84. Brouzes, E.; Medkova, M.; Savenelli, N.; Marran, D.; Twardowski, M.; Hutchison, J.B.; Rothberg, J.M.; Link, D.R.; Perrimon, N.; Samuels, M.L. Droplet microfluidic technology for single-cell high-throughput screening. *Proc. Natl. Acad. Sci. USA* **2009**, *106*, 14195–14200. [CrossRef]
85. Frenz, L.; Blank, K.; Brouzes, E.; Griffiths, A.D. Reliable microfluidic on-chip incubation of droplets in delay-lines. *Lab Chip* **2009**, *9*, 1344–1348. [CrossRef] [PubMed]
86. Adekanmbi, E.O.; Srivastava, S.K. Dielectrophoretic applications for disease diagnostics using lab-on-a-chip platforms. *Lab Chip* **2016**, *16*, 2148–2167. [CrossRef] [PubMed]
87. Chen, Y.; Li, P.; Huang, P.-H.; Xie, Y.; Mai, J.D.; Wang, L.; Nguyen, N.-T.; Huang, T.J. Rare cell isolation and analysis in microfluidics. *Lab Chip* **2014**, *14*, 626. [CrossRef] [PubMed]
88. Beech, J.P.; Jönsson, P.; Tegenfeldt, J.O. Tipping the balance of deterministic lateral displacement devices using dielectrophoresis. *Lab Chip* **2009**, *9*, 2698–2706. [CrossRef] [PubMed]
89. Beech, J.P.; Keim, K.; Ho, B.D.; Guiducci, C.; Tegenfeldt, J.O. Active Posts in Deterministic Lateral Displacement Devices. *Adv. Mater. Technol.* **2019**. [CrossRef]
90. Calero, V.; Garcia-Sanchez, P.; Honrado, C.; Ramos, A.; Morgan, H. AC electrokinetic biased deterministic lateral displacement for tunable particle separation. *Lab Chip* **2019**, *19*, 1386–1396. [CrossRef]
91. Smith, A.J.; O'Rorke, R.D.; Kale, A.; Rimsa, R.; Tomlinson, M.J.; Kirkham, J.; Davies, A.G.; Wälti, C.; Wood, C.D. Rapid cell separation with minimal manipulation for autologous cell therapies. *Sci. Rep.* **2017**, *7*, 41872. [CrossRef]
92. Sharifi Noghabi, H.; Soo, M.; Khamenehfar, A.; Li, P.C.H. Dielectrophoretic trapping of single leukemic cells using the conventional and compact optical measurement systems. *Electrophoresis* **2019**, *40*, 1478–1485. [CrossRef]
93. Faraghat, S.A.; Hoettges, K.F.; Steinbach, M.K.; van der Veen, D.R.; Brackenbury, W.J.; Henslee, E.A.; Labeed, F.H.; Hughes, M.P. High-throughput, low-loss, low-cost, and label-free cell separation using electrophysiology-activated cell enrichment. *Proc. Natl. Acad. Sci. USA* **2017**, *114*, 4591–4596. [CrossRef]
94. Huang, L.R.; Cox, E.C.; Austin, R.H.; Sturm, J.C. Continuous particle separation through deterministic lateral displacement. *Science* **2004**, *304*, 987–990. [CrossRef]
95. Liu, Z.; Huang, F.; Du, J.; Shu, W.; Feng, H.; Xu, X.; Chen, Y. Rapid isolation of cancer cells using microfluidic deterministic lateral displacement structure. *Biomicrofluidics* **2013**, *7*, 11801. [CrossRef] [PubMed]
96. Loutherback, K.; D'Silva, J.; Liu, L.; Wu, A.; Austin, R.H.; Sturm, J.C. Deterministic separation of cancer cells from blood at 10 mL/min. *AIP Adv.* **2012**, *2*, 42107. [CrossRef] [PubMed]

97. Davis, J.A.; Inglis, D.W.; Morton, K.J.; Lawrence, D.A.; Huang, L.R.; Chou, S.Y.; Sturm, J.C.; Austin, R.H. Deterministic hydrodynamics: Taking blood apart. *Proc. Natl. Acad. Sci. USA* **2006**, *103*, 14779–14784. [CrossRef] [PubMed]
98. Krüger, T.; Holmes, D.; Coveney, P.V. Deformability-based red blood cell separation in deterministic lateral displacement devices—A simulation study. *Biomicrofluidics* **2014**, *8*, 054114. [CrossRef]
99. Joensson, H.N.; Uhlén, M.; Svahn, H.A. Droplet size based separation by deterministic lateral displacement-separating droplets by cell-induced shrinking. *Lab Chip* **2011**, *11*, 1305–1310. [CrossRef]
100. Inglis, D.W.; Herman, N.; Vesey, G. Highly accurate deterministic lateral displacement device and its application to purification of fungal spores. *Biomicrofluidics* **2010**, *4*. [CrossRef]
101. Zeming, K.K.; Salafi, T.; Chen, C.H.; Zhang, Y. Asymmetrical Deterministic Lateral Displacement Gaps for Dual Functions of Enhanced Separation and Throughput of Red Blood Cells. *Sci. Rep.* **2016**, *6*, 22934. [CrossRef]
102. Wunsch, B.H.; Smith, J.T.; Gifford, S.M.; Wang, C.; Brink, M.; Bruce, R.L.; Austin, R.H.; Stolovitzky, G.; Astier, Y. Nanoscale lateral displacement arrays for the separation of exosomes and colloids down to 20 nm. *Nat. Nanotechnol.* **2016**, *11*, 936–940. [CrossRef]
103. Hou, S.; Chen, J.-F.; Song, M.; Zhu, Y.; Jan, Y.J.; Chen, S.H.; Weng, T.-H.; Ling, D.-A.; Chen, S.-F.; Ro, T.; et al. Imprinted NanoVelcro Microchips for Isolation and Characterization of Circulating Fetal Trophoblasts: Toward Noninvasive Prenatal Diagnostics. *ACS Nano* **2017**, *11*, 8167–8177. [CrossRef]
104. Pariset, E.; Pudda, C.; Boizot, F.; Verplanck, N.; Berthier, J.; Thuaire, A.; Agache, V. Anticipating Cutoff Diameters in Deterministic Lateral Displacement (DLD) Microfluidic Devices for an Optimized Particle Separation. *Small* **2017**, *13*. [CrossRef]
105. Vernekar, R.; Krüger, T. Breakdown of deterministic lateral displacement efficiency for non-dilute suspensions: A numerical study. *Med. Eng. Phys.* **2015**, *37*, 845–854. [CrossRef] [PubMed]
106. Holm, S.H.; Zhang, Z.; Beech, J.P.; Gompper, G.; Fedosov, D.A.; Tegenfeldt, J.O. Microfluidic Particle Sorting in Concentrated Erythrocyte Suspensions. *Phys. Rev. Appl.* **2019**, *12*, 014051. [CrossRef]
107. Alizadehrad, D.; Krüger, T.; Engstler, M.; Stark, H. Simulating the Complex Cell Design of Trypanosoma brucei and Its Motility. *PLoS Comput. Biol.* **2015**, *11*, e1003967. [CrossRef] [PubMed]
108. D'Avino, G. Non-Newtonian deterministic lateral displacement separator: Theory and simulations. *Rheol. Acta* **2013**, *52*, 221–236. [CrossRef]
109. Yan, S.; Zhang, J.; Yuan, D.; Li, W. Hybrid microfluidics combined with active and passive approaches for continuous cell separation. *Electrophoresis* **2017**, *38*, 238–249. [CrossRef]
110. Zhang, Z.; Chien, W.; Henry, E.; Fedosov, D.A.; Gompper, G. Sharp-edged geometric obstacles in microfluidics promote deformability-based sorting of cells. *Phys. Rev. Fluids* **2019**, *4*, 1–18. [CrossRef]
111. Imai, Y.; Kondo, H.; Ishikawa, T.; Teck Lim, C.; Yamaguchi, T. Modeling of hemodynamics arising from malaria infection. *J. Biomech.* **2010**, *43*, 1386–1393. [CrossRef]
112. Guo, Q.; Duffy, S.P.; Matthews, K.; Deng, X.; Santoso, A.T.; Islamzada, E.; Ma, H. Deformability based sorting of red blood cells improves diagnostic sensitivity for malaria caused by Plasmodium falciparum. *Lab Chip* **2016**, *16*, 645–654. [CrossRef]
113. Park, E.S.; Jin, C.; Guo, Q.; Ang, R.R.; Duffy, S.P.; Matthews, K.; Azad, A.; Abdi, H.; Todenhöfer, T.; Bazov, J.; et al. Continuous Flow Deformability-Based Separation of Circulating Tumor Cells Using Microfluidic Ratchets. *Small* **2016**, *12*, 1909–1919. [CrossRef]
114. Chen, H.; Zhang, Z.; Liu, H.; Zhang, Z.; Lin, C.; Wang, B. Hybrid magnetic and deformability based isolation of circulating tumor cells using microfluidics. *AIP Adv.* **2019**, *9*. [CrossRef]
115. Chen, H. A Triplet Parallelizing Spiral Microfluidic Chip for Continuous Separation of Tumor Cells. *Sci. Rep.* **2018**, *8*, 4042. [CrossRef] [PubMed]
116. Zhou, T.; Yeh, L.-H.; Li, F.-C.; Mauroy, B.; Joo, S. Deformability-Based Electrokinetic Particle Separation. *Micromachines* **2016**, *7*, 170. [CrossRef] [PubMed]
117. Hou, H.W.; Bhagat, A.A.S.; Lin Chong, A.G.; Mao, P.; Wei Tan, K.S.; Han, J.; Lim, C.T. Deformability based cell margination—A simple microfluidic design for malaria-infected erythrocyte separation. *Lab Chip* **2010**, *10*, 2605–2613. [CrossRef] [PubMed]
118. Imai, Y.; Nakaaki, K.; Kondo, H.; Ishikawa, T.; Teck Lim, C.; Yamaguchi, T. Margination of red blood cells infected by Plasmodium falciparum in a microvessel. *J. Biomech.* **2011**, *44*, 1553–1558. [CrossRef]

119. Xiang, N.; Wang, J.; Li, Q.; Han, Y.; Huang, D.; Ni, Z. Precise Size-Based Cell Separation via the Coupling of Inertial Microfluidics and Deterministic Lateral Displacement. *Anal. Chem.* **2019**, *91*, 10328–10334. [CrossRef]
120. Morsi, S.A.; Alexander, A.J. An investigation of particle trajectories in two-phase flow systems. *J. Fluid Mech.* **1972**, *55*, 193–208. [CrossRef]
121. Ho, B.P.; Leal, L.G. Inertial migration of rigid spheres in two-dimensional unidirectional flows. *J. Fluid Mech.* **1974**, *65*, 365–400. [CrossRef]
122. Liu, C.; Xue, C.; Sun, J.; Hu, G. A generalized formula for inertial lift on a sphere in microchannels. *Lab Chip* **2016**, *16*, 884–892. [CrossRef]
123. Zhang, J.; Yuan, D.; Zhao, Q.; Teo, A.J.T.; Yan, S.; Ooi, C.H.; Li, W.; Nguyen, N.-T. Fundamentals of Differential Particle Inertial Focusing in Symmetric Sinusoidal Microchannels. *Anal. Chem.* **2019**, *91*, 4077–4084. [CrossRef]
124. Sun, J.; Li, M.; Liu, C.; Zhang, Y.; Liu, D.; Liu, W.; Hu, G.; Jiang, X. Double spiral microchannel for label-free tumor cell separation and enrichment. *Lab Chip* **2012**, *12*, 3952–3960. [CrossRef]
125. Bhagat, A.A.S.; Kuntaegowdanahalli, S.S.; Papautsky, I. Continuous particle separation in spiral microchannels using dean flows and differential migration. *Lab Chip* **2008**, *8*, 1906–1914. [CrossRef]
126. Kemna, E.W.M.; Schoeman, R.M.; Wolbers, F.; Vermes, I.; Weitz, D.A.; Van Den Berg, A. High-yield cell ordering and deterministic cell-in-droplet encapsulation using Dean flow in a curved microchannel. *Lab Chip* **2012**, *12*, 2881–2887. [CrossRef] [PubMed]
127. Ookawara, S.; Street, D.; Ogawa, K. Numerical study on development of particle concentration profiles in a curved microchannel. *Chem. Eng. Sci.* **2006**, *61*, 3714–3724. [CrossRef]
128. Zuvin, M.; Mansur, N.; Birol, S.Z.; Trabzon, L.; Sayı Yazgan, A. Human breast cancer cell enrichment by Dean flow driven microfluidic channels. *Microsyst. Technol.* **2016**, *22*, 645–652. [CrossRef]
129. Zhou, J.; Giridhar, P.V.; Kasper, S.; Papautsky, I. Modulation of aspect ratio for complete separation in an inertial microfluidic channel. *Lab Chip* **2013**, *13*, 1919–1929. [CrossRef] [PubMed]
130. Mach, A.J.; di Carlo, D. Continuous scalable blood filtration device using inertial microfluidics. *Biotechnol. Bioeng.* **2010**, *107*, 302–311. [CrossRef]
131. Wu, Z.; Chen, Y.; Wang, M.; Chung, A.J. Continuous inertial microparticle and blood cell separation in straight channels with local microstructures. *Lab Chip* **2016**, *16*, 532–542. [CrossRef]
132. Shi, J.; Huang, H.; Stratton, Z.; Huang, Y.; Huang, T.J. Continuous particle separation in a microfluidic channel via standing surface acoustic waves (SSAW). *Lab Chip* **2009**, *9*, 3354–3359. [CrossRef]
133. Destgeer, G.; Lee, K.H.; Jung, J.H.; Alazzam, A.; Sung, H.J. Continuous separation of particles in a PDMS microfluidic channel via travelling surface acoustic waves (TSAW). *Lab Chip* **2013**, *13*, 4210–4216. [CrossRef]
134. Wiklund, M.; Green, R.; Ohlin, M. Acoustofluidics 14: Applications of acoustic streaming in microfluidic devices. *Lab Chip* **2012**, *12*, 2438–2451. [CrossRef]
135. Patel, M.V.; Tovar, A.R.; Lee, A.P. Lateral cavity acoustic transducer as an on-chip cell/particle microfluidic switch. *Lab Chip* **2012**, *12*, 139–145. [CrossRef]
136. Tovar, A.R.; Patel, M.V.; Lee, A.P. Lateral air cavities for microfluidic pumping with the use of acoustic energy. *Microfluid. Nanofluid.* **2011**, *10*, 1269–1278. [CrossRef]
137. Xu, Y.; Hashmi, A.; Yu, G.; Lu, X.; Kwon, H.J.; Chen, X.; Xu, J. Microbubble array for on-chip worm processing. *Appl. Phys. Lett.* **2013**, *102*. [CrossRef]
138. Ozcelik, A.; Rufo, J.; Guo, F.; Gu, Y.; Li, P.; Lata, J.; Huang, T.J. Acoustic tweezers for the life sciences. *Nat. Methods* **2018**, *15*, 1021–1028. [CrossRef]
139. Ashkin, A.; Dziedzic, J.M. Optical Levitation by Radiation Pressure. *Appl. Phys. Lett.* **1971**, *19*, 283. [CrossRef]
140. Flynn, R.A.; Shao, B.; Chachisvilis, M.; Ozkan, M.; Esener, S.C. Two-beam optical traps: Refractive index and size measurements of microscale objects. *Biomed. Microdevices* **2005**, *7*, 93–97. [CrossRef]
141. Berg-Sørensen, K.; Marie, R.; Dziegiel, M.H.; Nielsen, K.S.; Rungling, T.B. Deformation of Single Cells—Optical Two-Beam Traps and More. In *International Society for Optics and Photonics, Proceedings of the Complex Light and Optical Forces XIII, San Francisco, CA, USA, 2–7 February 2019*; Andrews, D.L., Galvez, E.J., Glückstad, J., Eds.; SPIE: Bellingham, WA, USA, 2019; Volume 10935, p. 39.
142. Zhang, H.; Liu, K.-K. Optical tweezers for single cells. *J. R. Soc. Interface* **2008**, *5*, 671–690. [CrossRef]
143. Liu, Y.; Yu, M. Investigation of inclined dual-fiber optical tweezers for 3D manipulation and force sensing. *Opt. Express* **2009**, *17*, 13624–13638. [CrossRef]
144. Grier, D.G. A revolution in optical manipulation. *Nature* **2003**, *424*, 810–816. [CrossRef]

145. Padgett, M.; Di Leonardo, R. Holographic optical tweezers and their relevance to lab on chip devices. *Lab Chip* **2011**, *11*, 1196–1205. [CrossRef]
146. Santucci, S.C.; Cojoc, D.; Amenitsch, H.; Marmiroli, B.; Sartori, B.; Burghammer, M.; Schoeder, S.; DiCola, E.; Reynolds, M.; Riekel, C. Optical tweezers for synchrotron radiation probing of trapped biological and soft matter objects in aqueous environments. *Anal. Chem.* **2011**, *83*, 4863–4870. [CrossRef] [PubMed]
147. Schonbrun, E.; Piestun, R.; Jordan, P.; Cooper, J.; Wulff, K.D.; Courtial, J.; Padgett, M. 3D interferometric optical tweezers using a single spatial light modulator. *Opt. Express* **2005**, *13*, 3777. [CrossRef] [PubMed]
148. Sinclair, G.; Jordan, P.; Courtial, J.; Padgett, M.; Cooper, J.; Laczik, Z. Assembly of 3-dimensional structures using programmable holographic optical tweezers. *Opt. Express* **2004**, *12*, 5475–5480. [CrossRef]
149. Leach, J.; Sinclair, G.; Jordan, P.; Courtial, J.; Padgett, M.; Cooper, J.; Laczik, Z. 3D manipulation of particles into crystal structures using holographic optical tweezers. *Opt. Express* **2004**, *12*, 220–226. [CrossRef] [PubMed]
150. Preece, D.; Keen, S.; Botvinick, E.; Bowman, R.; Padgett, M.; Leach, J. Independent polarisation control of multiple optical traps. *Opt. Express* **2008**, *16*, 15897. [CrossRef]
151. Hesseling, C.; Woerdemann, M.; Hermerschmidt, A.; Denz, C. Controlling ghost traps in holographic optical tweezers. *Opt. Lett.* **2011**, *36*, 3657–3659. [CrossRef]
152. Dalal, A.; Chowdhury, A.; Dasgupta, R.; Majumder, S.K. Improved generation of periodic optical trap arrays using noniterative algorithm. *Opt. Eng.* **2017**, *56*. [CrossRef]
153. Tang, X.; Nan, F.; Han, F.; Yan, Z. Simultaneously shaping the intensity and phase of light for optical nanomanipulation. *arXiv* **2019**, arXiv:1910.08244.
154. Perozziello, G.; Candeloro, P.; Laura Coluccio, M.; Di Fabrizio, E. Optofluidics for handling and analysis of single living cells. *Nanofluid* **2017**, *4*, 18–23. [CrossRef]
155. Jorgolli, M.; Nevill, T.; Winters, A.; Chen, I.; Chong, S.; Lin, F.F.; Mock, M.; Chen, C.; Le, K.; Tan, C.; et al. Nanoscale integration of single cell biologics discovery processes using optofluidic manipulation and monitoring. *Biotechnol. Bioeng.* **2019**, *116*, 2393–2411. [CrossRef]
156. Li, X.; Yang, H.; Huang, H.; Sun, D. A switching controller for high speed cell transportation by using a robot-aided optical tweezers system. *Automatica* **2018**, *89*, 308–315. [CrossRef]
157. Xie, M.; Shakoor, A.; Wu, C. Manipulation of Biological Cells Using a Robot-Aided Optical Tweezers System. *Micromachines* **2018**, *9*, 245. [CrossRef] [PubMed]
158. Carey, T.R.; Cotner, K.L.; Li, B.; Sohn, L.L. Developments in label-free microfluidic methods for single-cell analysis and sorting. *Wiley Interdiscip. Rev. Nanomed. Nanobiotechnol.* **2019**, *11*, e1529. [CrossRef]
159. Wang, M.M.; Tu, E.; Raymond, D.E.; Yang, J.M.; Zhang, H.; Hagen, N.; Dees, B.; Mercer, E.M.; Forster, A.H.; Kariv, I.; et al. Microfluidic sorting of mammalian cells by optical force switching. *Nat. Biotechnol.* **2005**, *23*, 83–87. [CrossRef]
160. Sajeesh, P.; Sen, A.K. Particle separation and sorting in microfluidic devices: A review. *Microfluid. Nanofluidics* **2014**, *17*, 1–52. [CrossRef]
161. Cao, B.; Kelbauskas, L.; Chan, S.; Shetty, R.M.; Smith, D.; Meldrum, D.R. Rotation of single live mammalian cells using dynamic holographic optical tweezers. *Opt. Lasers Eng.* **2017**, *92*, 70–75.
162. Keloth, A.; Anderson, O.; Risbridger, D.; Paterson, L. Single cell isolation using optical tweezers. *Micromachines* **2018**, *9*, 434.
163. Rappaz, B.; Marquet, P.; Cuche, E.; Emery, Y.; Depeursinge, C.; Magistretti, P.J. Measurement of the integral refractive index and dynamic cell morphometry of living cells with digital holographic microscopy. *Opt. Express* **2005**, *13*, 9361. [CrossRef]
164. Curl, C.L.; Bellair, C.J.; Harris, T.; Allman, B.E.; Harris, P.J.; Stewart, A.G.; Roberts, A.; Nugent, K.A.; Delbridge, L.M.D. Refractive index measurement in viable cells using quantitative phase-amplitude microscopy and confocal microscopy. *Cytom. Part A* **2005**, *65*, 88–92. [CrossRef]
165. Kaminski, T.S.; Garstecki, P. Controlled droplet microfluidic systems for multistep chemical and biological assays. *Chem. Soc. Rev.* **2017**, *46*, 6210–6226. [CrossRef]
166. Liu, C.; Xu, X.; Li, B.; Situ, B.; Pan, W.; Hu, Y.; An, T.; Yao, S.; Zheng, L. Single-Exosome-Counting Immunoassays for Cancer Diagnostics. *Nano Lett.* **2018**, *18*, 4226–4232. [CrossRef] [PubMed]
167. Fornell, A.; Cushing, K.; Nilsson, J.; Tenje, M. Binary particle separation in droplet microfluidics using acoustophoresis. *Appl. Phys. Lett.* **2018**, *112*. [CrossRef]

168. Huh, D.; Bahng, J.H.; Ling, Y.; Wei, H.-H.; Kripfgans, O.D.; Fowlkes, J.B.; Grotberg, J.B.; Takayama, S. Gravity-Driven Microfluidic Particle Sorting Device with Hydrodynamic Separation Amplification. *Anal. Chem.* **2007**, *79*, 1369–1376. [CrossRef] [PubMed]
169. Seemann, R.; Brinkmann, M.; Pfohl, T.; Herminghaus, S. Droplet based microfluidics. *Rep. Prog. Phys.* **2012**, *75*, 016601. [CrossRef]
170. Chabert, M.; Viovy, J.-L. Microfluidic high-throughput encapsulation and hydrodynamic self-sorting of single cells. *Proc. Natl. Acad. Sci. USA* **2008**, *105*, 3191–3196. [CrossRef]
171. Tan, Y.C.; Ho, Y.L.; Lee, A.P. Droplet coalescence by geometrically mediated flow in microfluidic channels. *Microfluid. Nanofluid.* **2007**, *3*, 495–499. [CrossRef]
172. Fidalgo, L.M.; Whyte, G.; Bratton, D.; Kaminski, C.F.; Abell, C.; Huck, W.T.S. From Microdroplets to Microfluidics: Selective Emulsion Separation in Microfluidic Devices. *Angew. Chem. Int. Ed.* **2008**, *47*, 2042–2045. [CrossRef]
173. Ding, R.; Lloyd Ung, W.; Heyman, J.A.; Weitz, D.A. Sensitive and predictable separation of microfluidic droplets by size using in-line passive filter. *Biomicrofluidics* **2017**, *11*, 014114. [CrossRef]
174. Kim, S.; Ganapathysubramanian, B.; Anand, R.K. Concentration Enrichment, Separation, and Cation Exchange in Nanoliter-Scale Water-in-Oil Droplets. *J. Am. Chem. Soc.* **2020**, *142*, 3196–3204. [CrossRef]
175. Huang, C.; Han, S.I.; Han, A. In-Droplet Cell Separation Based on Different Dielectrophoretic Response. In Proceedings of the 2019 20th International Conference on Solid-State Sensors, Actuators and Microsystems and Eurosensors XXXIII, Transducers 2019 and Eurosensors XXXIII, Berlin, Germany, 23–27 June 2019; pp. 492–495.
176. Wang, Y.; Zhao, Y.; Cho, S.K. Efficient in-droplet separation of magnetic particles for digital microfluidics. *J. Micromech. Microeng.* **2007**, *17*, 2148–2156. [CrossRef]
177. Brouzes, E.; Kruse, T.; Kimmerling, R.; Strey, H.H. Rapid and continuous magnetic separation in droplet microfluidic devices. *Lab Chip* **2015**, *15*, 908–919. [CrossRef] [PubMed]
178. Wang, S.; Sung, K.-J.; Lin, X.N.; Burns, M.A. Bead mediated separation of microparticles in droplets. *PLoS ONE* **2017**, *12*, e0173479. [CrossRef] [PubMed]
179. Sista, R.S.; Ng, R.; Nuffer, M.; Basmajian, M.; Coyne, J.; Elderbroom, J.; Hull, D.; Kay, K.; Krishnamurthy, M.; Roberts, C.; et al. Digital Microfluidic Platform to Maximize Diagnostic Tests with Low Sample Volumes from Newborns and Pediatric Patients. *Diagnostics* **2020**, *10*, 21. [CrossRef] [PubMed]
180. Liang, X.J.; Liu, A.Q.; Lim, C.S.; Ayi, T.C.; Yap, P.H. Determining refractive index of single living cell using an integrated microchip. *Sens. Actuators A Phys.* **2007**, *133*, 349–354. [CrossRef]
181. Liu, P.Y.; Chin, L.K.; Ser, W.; Ayi, T.C.; Yap, P.H.; Bourouina, T.; Leprince-Wang, Y. An optofluidic imaging system to measure the biophysical signature of single waterborne bacteria. *Lab Chip* **2014**, *14*, 4237–4243. [CrossRef] [PubMed]
182. Song, W.Z.; Zhang, X.M.; Liu, A.Q.; Lim, C.S.; Yap, P.H.; Hosseini, H.M.M. Refractive index measurement of single living cells using on-chip Fabry-Prot cavity. *Appl. Phys. Lett.* **2006**, *89*, 15–18. [CrossRef]
183. Song, W.Z.; Liu, A.Q.; Swaminathan, S.; Lim, C.S.; Yap, P.H.; Ayi, T.C. Determination of single living cell's dry/water mass using optofluidic chip. *Appl. Phys. Lett.* **2007**, *91*, 1–4. [CrossRef]
184. Chin, L.K.; Liu, A.Q.; Lim, C.S.; Zhang, X.M.; Ng, J.H.; Hao, J.Z.; Takahashi, S. Differential single living cell refractometry using grating resonant cavity with optical trap. *Appl. Phys. Lett.* **2007**, *91*, 1–4. [CrossRef]
185. Liu, P.Y.; Chin, L.K.; Ser, W.; Chen, H.F.; Hsieh, C.M.; Lee, C.H.; Sung, K.B.; Ayi, T.C.; Yap, P.H.; Liedberg, B.; et al. Cell refractive index for cell biology and disease diagnosis: Past, present and future. *Lab Chip* **2016**, *16*, 634–644. [CrossRef]
186. Prevedel, R.; Diz-Muñoz, A.; Ruocco, G.; Antonacci, G. Brillouin microscopy—A revolutionary tool for mechanobiology? *arXiv* **2019**, arXiv:1901.02006.
187. Bateman, J.B.; Wagman, J.; Carstensen, E.L. Refraction and absorption of light in bacterial suspensions. *Kolloid-Zeitschrift Zeitschrift für Polymere* **1966**, *208*, 44–58. [CrossRef]
188. Balaev, A.E.; Dvoretski, K.N.; Doubrovski, V.A. Refractive Index of Escherichia Coli Cells. In Proceedings of the SPIE 4707, Saratov Fall Meeting 2001: Optical Technologies in Biophysics and Medicine III, Saratov, Russia, 16 July 2002.
189. Tuminello, P.S.; Arakawa, E.T.; Khare, B.N.; Wrobel, J.M.; Querry, M.R.; Milham, M.E. Optical properties of Bacillus subtilis spores from 02 to 25 µm. *Appl. Opt.* **1997**, *36*, 2818. [CrossRef]
190. Joseph, S. Refractometry of fungi. *J. Microsc.* **1983**, *131*, 163–172. [CrossRef]

191. Hsu, W.-C.; Su, J.-W.; Tseng, T.-Y.; Sung, K.-B. Tomographic diffractive microscopy of living cells based on a common-path configuration. *Opt. Lett.* **2014**, *39*, 2210. [CrossRef]
192. Charrière, F.; Marian, A.; Montfort, F.; Kuehn, J.; Colomb, T.; Cuche, E.; Marquet, P.; Depeursinge, C. Cell refractive index tomography by digital holographic microscopy. *Opt. Lett.* **2006**, *31*, 178. [CrossRef] [PubMed]
193. Lue, N.; Popescu, G.; Ikeda, T.; Badizadegan, K.; Dasari, R.R.; Feld, M.S. Live cell refractometry using hilbert phase microscopy. *Opt. InfoBase Conf. Pap.* **2006**, *113*, 13327–13330.
194. Sung, Y.; Lue, N.; Hamza, B.; Martel, J.; Irimia, D.; Dasari, R.R.; Choi, W.; Yaqoob, Z.; So, P. Three-dimensional holographic refractive-index measurement of continuously flowing cells in a microfluidic channel. *Phys. Rev. Appl.* **2014**, *1*, 1–8.
195. Su, J.W.; Hsu, W.C.; Chou, C.Y.; Chang, C.H.; Sung, K. Bin Digital holographic microtomography for high-resolution refractive index mapping of live cells. *J. Biophotonics* **2013**, *6*, 416–424. [CrossRef]
196. Chu, Y.C.; Chang, W.Y.; Chen, K.H.; Chen, J.H.; Tsai, B.C.; Hsu, K.Y. Full-field refractive index measurement with simultaneous phase-shift interferometry. *Optik* **2014**, *125*, 3307–3310. [CrossRef]
197. Moh, K.J.; Yuan, X.-C.; Bu, J.; Zhu, S.W.; Gao, B.Z. Surface plasmon resonance imaging of cell-substrate contacts with radially polarized beams. *Opt. Express* **2008**, *16*, 20734. [CrossRef]
198. Lee, J.Y.; Lee, C.W.; Lin, E.H.; Wei, P.K. Single live cell refractometer using nanoparticle coated fiber tip. *Appl. Phys. Lett.* **2008**, *93*, 173110–173113. [CrossRef]
199. Phillips, K.G.; Jacques, S.L.; McCarty, O.J.T. Measurement of Single Cell Refractive Index, Dry Mass, Volume, and Density Using a Transillumination Microscope. *Phys. Rev. Lett.* **2012**, *109*, 1–5. [CrossRef] [PubMed]
200. Martinez, A.W.; Phillips, S.T.; Butte, M.J.; Whitesides, G.M. Patterned Paper as a Platform for Inexpensive, Low-Volume, Portable Bioassays. *Angew. Chem. Int. Ed.* **2007**, *46*, 1318–1320. [CrossRef] [PubMed]
201. Nilghaz, A.; Guan, L.; Tan, W.; Shen, W. Advances of Paper-Based Microfluidics for Diagnostics—The Original Motivation and Current Status. *ACS Sens.* **2016**, *1*, 1382–1393. [CrossRef]
202. He, Y.; Wu, Y.; Fu, J.-Z.; Wu, W.-B. Fabrication of paper-based microfluidic analysis devices: A review. *RSC Adv.* **2015**, *5*, 78109–78127. [CrossRef]
203. Martinez, A.W.; Phillips, S.T.; Carrilho, E.; Thomas, S.W.; Sindi, H.; Whitesides, G.M. Simple Telemedicine for Developing Regions: Camera Phones and Paper-Based Microfluidic Devices for Real-Time, Off-Site Diagnosis. *Anal. Chem.* **2008**, *80*, 3699–3707. [CrossRef]
204. Lisowski, P.; Zarzycki, P.K. Microfluidic paper-based analytical devices (μPADs) and micro total analysis systems (μTAS): Development, applications and future trends. *Chromatographia* **2013**, *76*, 1201–1214. [CrossRef]
205. Su, W.; Gao, X.; Jiang, L.; Qin, J. Microfluidic platform towards point-of-care diagnostics in infectious diseases. *J. Chromatogr. A* **2015**, *1377*, 13–26. [CrossRef]
206. Mu, X.; Zhang, L.; Chang, S.; Cui, W.; Zheng, Z. Multiplex microfluidic paper-based immunoassay for the diagnosis of hepatitis C virus infection. *Anal. Chem.* **2014**, *86*, 5338–5344. [CrossRef]
207. Thom, N.K.; Yeung, K.; Pillion, M.B.; Phillips, S.T. "Fluidic batteries" as low-cost sources of power in paper-based microfluidic devices. *Lab Chip* **2012**, *12*, 1768–1770. [CrossRef]
208. Sánchez-Ovejero, C.; Benito-Lopez, F.; Díez, P.; Casulli, A.; Siles-Lucas, M.; Fuentes, M.; Manzano-Román, R. Sensing parasites: Proteomic and advanced bio-detection alternatives. *J. Proteomics* **2016**, *136*, 145–156. [CrossRef] [PubMed]
209. Zole, E.; Ranka, R. Mitochondria, its DNA and telomeres in ageing and human population. *Biogerontology* **2018**, *19*, 189–208. [CrossRef] [PubMed]
210. Ma, S.Y.; Chiang, Y.C.; Hsu, C.H.; Chen, J.J.; Hsu, C.C.; Chao, A.C.; Lin, Y.S. Peanut detection using droplet microfluidic polymerase chain reaction device. *J. Sens.* **2019**, *2019*, 1–9. [CrossRef]
211. Gascoyne, P.; Satayavivad, J.; Ruchirawat, M. Microfluidic approaches to malaria detection. *Acta Trop.* **2004**, *89*, 357–369. [CrossRef]
212. Ryan, U.; Paparini, A.; Oskam, C. New Technologies for Detection of Enteric Parasites. *Trends Parasitol.* **2017**, *33*, 532–546. [CrossRef]
213. Ahrberg, C.D.; Manz, A.; Chung, B.G. Polymerase chain reaction in microfluidic devices. *Lab Chip* **2016**, *16*, 3866–3884. [CrossRef]
214. Notomi, T.; Okayama, H.; Masubuchi, H.; Yonekawa, T.; Watanabe, K.; Amino, N.; Hase, T. Loop-Mediated Isothermal Amplification of DNA. *Nucleic Acids Res.* **2000**, *28*, e63. [CrossRef]

215. Haring, R.; Wallaschofski, H. Diving through the "-omics": The case for deep phenotyping and systems epidemiology. *OMICS* **2012**, *16*, 231–234. [CrossRef]
216. Nassar, S.; Raddassi, K.; Wu, T.; Wisnewski, A. Single-Cell Metabolomics by Mass Spectrometry for Drug Discovery: Moving Forward. *Drug Des. Open Access* **2017**, *6*, 1000149.
217. Wang, D.; Bodovitz, S. Single cell analysis: The new frontier in "omics". *Trends Biotechnol.* **2010**, *28*, 281–290. [CrossRef]
218. Ogbaga, C.C.; Stepien, P.; Dyson, B.C.; Rattray, N.J.W.; Ellis, D.I.; Goodacre, R.; Johnson, G.N. Biochemical Analyses of Sorghum Varieties Reveal Differential Responses to Drought. *PLoS ONE* **2016**. [CrossRef]
219. Gika, H.G.; Theodoridis, G.A.; Wingate, J.E.; Wilson, I.D. Within-day reproducibility of an HPLC-MS-based method for metabonomic analysis: Application to human urine. *J. Proteome Res.* **2007**, *6*, 3291–3303. [CrossRef]
220. Soga, T.; Ohashi, Y.; Ueno, Y.; Naraoka, H.; Tomita, M.; Nishioka, T. Quantitative Metabolome Analysis Using Capillary Electrophoresis Mass Spectrometry. *J. Proteome Res.* **2003**, *2*, 488–494. [CrossRef]
221. Bissonnette, L.; Bergeron, M.G. *POC Tests in Microbial Diagnostics: Current Status. Methods in Microbiology*, 1st ed.; Elsevier Ltd.: Amsterdam, The Netherlands, 2015; Volume 42.
222. Tay, A.; Pavesi, A.; Yazdi, S.R.; Lim, C.T.; Warkiani, M.E. Advances in microfluidics in combating infectious diseases. *Biotechnol. Adv.* **2016**, *34*, 404–421. [CrossRef]
223. Sharma, S.; Zapatero-Rodríguez, J.; Estrela, P.; O'Kennedy, R. Point-of-Care Diagnostics in Low Resource Settings: Present Status and Future Role of Microfluidics. *Biosensors* **2015**, *5*, 577–601. [CrossRef]
224. Zhang, Y.S.; Trujillo-de Santiago, G.; Alvarez, M.M.; Schiff, S.J.; Boyden, E.S.; Khademhosseini, A. Expansion mini-microscopy: An enabling alternative in point-of-care diagnostics. *Curr. Opin. Biomed. Eng.* **2017**, *1*, 45–53. [CrossRef]
225. Chaturvedi, A.; Gorthi, S.S. Automated blood sample preparation unit (ABSPU) for portable microfluidic flow cytometry. *SLAS Technol.* **2017**, *22*, 73–80. [CrossRef]
226. Yager, P.; Edwards, T.; Fu, E.; Helton, K.; Nelson, K.; Tam, M.R.; Weigl, B.H. Microfluidic diagnostic technologies for global public health. *Nature* **2006**, *442*, 412–418. [CrossRef]

© 2020 by the author. Licensee MDPI, Basel, Switzerland. This article is an open access article distributed under the terms and conditions of the Creative Commons Attribution (CC BY) license (http://creativecommons.org/licenses/by/4.0/).

Article

On-Chip Multiple Particle Velocity and Size Measurement Using Single-Shot Two-Wavelength Differential Image Analysis

Shuya Sawa [1], Mitsuru Sentoku [1] and Kenji Yasuda [1,2,*]

1 Department of Physics, School of Advanced Science and Engineering, Waseda University, Tokyo 169-8555, Japan; tkdc10@moegi.waseda.jp (S.S.); mitsen1019@fuji.waseda.jp (M.S.)
2 Department of Pure and Applied Physics, Graduate School of Advanced Science and Engineering, Waseda University, Tokyo 169-8555, Japan
* Correspondence: yasuda@list.waseda.jp

Received: 1 November 2020; Accepted: 14 November 2020; Published: 17 November 2020

Abstract: Precise and quick measurement of samples' flow velocities is essential for cell sorting timing control and reconstruction of acquired image-analyzed data. We developed a simple technique for the single-shot measurement of flow velocities of particles simultaneously in a microfluidic pathway. The speed was calculated from the difference in the particles' elongation in an acquired image that appeared when two wavelengths of light with different irradiation times were applied. We ran microparticles through an imaging flow cytometer and irradiated two wavelengths of light with different irradiation times simultaneously to those particles. The mixture of the two wavelength transmitted lights was divided into two wavelengths, and the images of the same microparticles for each wavelength were acquired in a single shot. We estimated the velocity from the difference of its elongation divided by the difference of irradiation time by comparing these two images. The distribution of polystyrene beads' velocity was parabolic and highest at the center of the flow channel, consistent with the expected velocity distribution of the laminar flow. Applying the calculated velocity, we also restored the accurate shapes and cross-sectional areas of particles in the images, indicating this simple method for improving of imaging flow cytometry and cell sorter for diagnostic screening of circulating tumor cells.

Keywords: imaging flow cytometer; precise velocity measurement; single-shot image-based velocity measurement; particle shape reconstruction; multi-view imaging; exposure time difference

1. Introduction

Numerous microfluidic devices have been used for cellular analysis technologies in biological research and medical diagnoses [1–8]. Recently, the change of cells' morphological characteristics as they pass through the cell cycle was used to sort particular yeasts in a fully automated low-cost imaging cell sorter with a cell detection rate of 5 s [9]. A low-cost microfluidic eight-way simultaneous image data-based sorting system mounted on a fluorescence microscope was also demonstrated with a sorting speed of 0.5 s resolution [10]. Furthermore, an automated imaging cell sorter system for cell identification within a flowing droplet was recently reported [11]. For the improvement of image analysis abilities, a deep-learning-assisted imaging cell sorter system was documented [12], in which the abilities of image analysis and cell identification were significantly improved while maintaining a time resolution on the sub-second order. Another unique machine learning-based approach for identifying of target cells with direct diffraction detection and analysis without any cell image acquisition was proposed in [13].

Image and size information of cells were applied for detection and identification of target cells for diagnostic applications [3,14–17], and recently, three-dimensional (3-D) images of single cells and their clusters were acquired [18]. However, the challenges of establishing a precise high-throughput for those technologies remained an issue and must be improved. Hence, the precise high-throughput image cell sorter combining with those existing methods can give us more flexible new approaches to medical diagnosis. For example, in our previous studies, we proposed an original approach for circulating tumor cell (CTC) detection in blood based on analyses of the morphometric parameters of cells without any antibody staining, with these parameters acting as "imaging biomarkers" [19–22]. For this approach, we developed a high-speed cell image recognition and analysis system to identify the target cells using their morphological characteristics during their flow within a narrow microfluidic pathway for sorting. We also improved this system and developed an on-chip multi-imaging flow cytometry system [23–25], which allows the simultaneous acquisition of both bright-field (BF) and fluorescent (FL) images when scanning all cells in the blood. Using this system, clustered CTCs were accurately detected for identifying their cluster sizes and their nucleus shapes and numbers within clusters simultaneously in the blood of cancer cell-implanted rats.

In contrast to the significant advancement in image recognition and analysis technologies in the field of imaging cell sorting, the fundamental and critical technical issues of the cell sorting process, such as the precise measurement of microparticle flow speed for correct target collection, has had no significant progress. For example, in our previous studies, only the mean values of the estimated flow velocities were applied for sorting timing set-up. Moreover, the influence of the exposure time of image acquisition, which is the origin of image blur, was neglected for the precise measurement of the acquired shapes of targets or their size estimations; we desired a method to overcome this limitation for more accurate and strict measurement and sorting of target samples.

We report the ability and limitation of our newly developed particle flow velocity measurement method with two-wavelength simultaneous one-shot differential image analysis. The results showed that this method could acquire the velocities of all of the spread particles in the microfluidic flow simultaneously only with a one-shot of image acquisition. In contrast, the conventional measurement was based on the two-shot image comparison for single target samples. The acquired particle velocities were also applied for the correction of the cross-sectional areas of particles, which were elongated by their movement during the exposure time of image acquisition. The ability and limitation of this method caused by shutter-time and image pixel size resolution of hardware for this analysis method were also discussed.

2. Principle

First, we explain the principle of the single-shot simultaneous two-wavelength differential imaging analysis for precise particle flow velocity measurement. In this method, we exploited the difference of exposure time of two light sources for image acquisition effectively for calculating the flow velocities of particles within the microfluidic pathway by measuring the difference of elongation length of the two images of the same particle by only a single shot of image acquisition interval. Usually, the exposure time is necessary to acquire the image and is also the reason for image blur, which is caused by the movement of targets within the exposure time. However, in this method, we used this image blur effectively to determine the precise flow velocities of particles. When we irradiate two different wavelength of lights with different lengths of time to the flowing particle, the acquired length of the particle should be elongated to the direction of flow depending on the exposure time of image acquisition. Hence, we can compare the length of the particle in two images of different wavelengths and can estimate the velocity by the difference of irradiation time of two-wavelength lights. Therefore, in this method, we only need one shot of image acquisition for velocity analysis and can apply this method to all the particles within a picture simultaneously without any complicated procedure of analysis. The advantages of this method are (1) only a single shot of image acquisition interval is needed for the estimation of flow velocity; (2) the velocities of all the particles within an image can be analyzed simultaneously; (3) the ability to reconstruct the actual image data from the accurate flow velocities to overcome the influence of image blur; (4) the system is easily set-up (only two different wavelengths, the exposure time of light source, and image dividing unit are needed).

Figure 1 is a schematic drawing explaining the principle of the method applied for our experiments as an example. Figure 1a illustrates the time chart of the camera's image acquisition interval, electronic shutter, externally inserted halogen lamp with a band-pass filter (590 nm), and a light emitting diode pulse (LED) signal source (530 nm). The frame rate of the Charge-Coupled Device (CCD) camera is 200 frames per second (fps), so one image is acquired every 5000 µs. The camera's electronic shutter, adjustable between 2 and 4994 µs, limits the exposure time of the halogen lamp that continues to irradiate. As a second light source, an LED with a period of 5000 µs (irradiation time: 2500 µs) synchronized with the image acquisition timing is irradiated.

Figure 1b illustrates the principle of this method to acquire the actual images of moving target samples: First, the image of the round shape sample (i) is elongated by the flow (movement) during an interval of image capture frame (ii), (iii). Simultaneous irradiation of two different wavelengths at irradiation times t_1 and t_2 ($t_1 > t_2$) causes a difference in lengths (L_1, L_2), respectively. The particle's velocity can be calculated from the difference in lengths and irradiation times. The elongation of the particle image is determined by multiplying the calculated speed and irradiation time. The accurate shaped particles' image is acquired by image processing, where the lower part of the elongated particle is lifted by the amount of the elongation (iv). On the other hand, it is impossible to restore the correct shape with the conventional image processing method that solely considers the vertical scale and compresses the flow direction (Y-axis) equally (v).

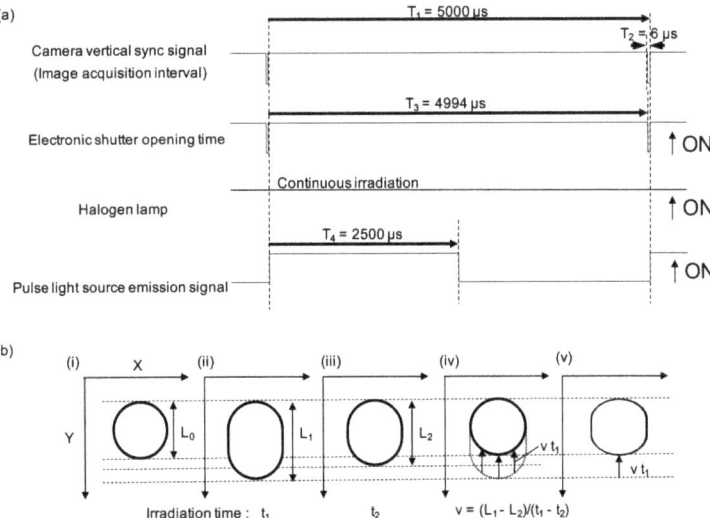

Figure 1. Restoration principle of accurate particle shape. (**a**) The time scale of the camera's image acquisition interval, electronic shutter, externally inserted halogen lamp, and a light emitting diode (LED) pulsed light signal. (**b**) Particle images acquired by photographing with different irradiation times with two restoration principles. (i) Image of a stationary particle. (ii), (iii) Images of particles moving in the Y-axis direction taken at irradiation time t_1 and t_2 ($t_1 > t_2$), respectively. The different irradiation times cause the difference in lengths (L_1, L_2). The velocity of moving particles can be calculated from the difference in irradiation time and lengths. (iv) Image restoration processing of the accurate particle shape. The lower part of the elongated particle is lifted by the amount of elongation determined by multiplying the calculated velocity and irradiation time. (v) The shape of the particle is restored by the conventional restoration principle that presses the Y-direction for the amount of elongation.

3. Materials and Methods

3.1. Measurement System

The measurement system consists of seven main parts: two light sources, a dichroic mirror unit for light mixing, a microfluidic chip (microchip), an objective lens, a multi-view module to divide an image into the two images of different wavelengths, a CCD camera camera, and a computer. A halogen lamp (LG-PS2; Olympus Co., Tokyo, Japan) was used for the 590 nm light source with 590 nm band-pass filter. An LED light (LED4D069, THORLABS, Newton, NJ, USA) was used for another 530 nm light source. The magnitude of the objective lens (x20, UPLSAPO, Olympus Co., Tokyo, Japan) was chosen to meet the microchannel width of the microchip. The multi-view module was custom made having a two-wavelength image dividing function. A high-speed digital CCD camera (HXC13; Baumer, Friedberg, Germany) was used for every 5000 μs image acquisition.

3.2. Microfluidic Chip

A disposable microfluidic chip (microchip) was fabricated with polydimethylsiloxane (PDMS) (SYLGARD 184 silicon elastomer; Dow Corning Co., Midland, MI, USA) attached to a thin glass slide by the same procedure as we reported previously [23], and was used for monitoring. In the microchip, the upper stream was divided into three channels: the central one was used for the sample inlet, and the remaining two side channels were used for sheath buffer inlets. After the junction at which the sample and sheath flow meet, the width of the sample flow was focused by hydrodynamic focusing, which allowed imaging of every single particle upon the arrangement of all of the particles in a straight line. The hydrodynamic focusing in this chip design conferred several advantages, such as centering the samples in the microfluidic flow, adding spaces between neighboring samples when lining them up, and aligning the orientation of samples in the direction of flow.

3.3. Preparation of Samples

In this experiment, two types of samples were used. Micro polyethylene spheres (diameter was 12 μm, Thermo Fisher Scientific K.K., Tokyo, Japan) were used as the standard size microparticle. They were diluted with pure water (6.38×10^5 beads/mL) and injected into the sample inlet of the microfluidic chip. Human cervix epithelioid carcinoma HeLa cells (ATCC, Manassas, VA, USA) were used as the standard cell sample. They were cultured in Dulbecco's Modified Eagle Medium (DMEM, Thermo Fisher Scientific K.K., Tokyo, Japan) with 10% Fetal Bovine Serum (FBS, Thermo Fisher Scientific K.K., Tokyo, Japan) and 1% Penicillin-Streptomycin (Thermo Fisher Scientific K.K., Tokyo, Japan). They were adjusted with culture medium (6.00×10^4 cells/mL) and injected into the sample inlet.

3.4. Lithography Processing with SU-8

Micropatterns designed by CAD were drawn on the mask blanks with a laser lithography system. Then, a photoresist layer and chromium layer were removed. After the SU-8 (SU-8 3000, KAYAKU Co. Ltd., Tokyo, Japan) coating process, the substrate was soft-baked to evaporate the solvent. Soft baking temperature and duration were 95 °C and 5 min, respectively. The substrate with a photomask was then exposed to irradiation under a UV lamp to induce cross-linkage. After the irradiation process, a Post-Exposure Bake (PEB) was performed. Post exposure baking had two steps. At the first step, temperature and duration were 65 °C and 1 min. At the second step, temperature and time were 95 °C and 5 min. Ultimately, the SU-8 resist that remained unhardened was removed by the SU-8 developer.

3.5. Lithography Processing with PDMS

Liquid PDMS was placed in the middle of the SU-8 mold and applied pressure for 15 min to uniformly spread the PDMS from the center to the whole SU-8 pattern. To harden the applied PDMS, the SU-8 mold was baked at 65 °C for 1 h. After the PDMS substrate became hardened, the PDMS was peeled from the mold carefully.

3.6. Program Construction of Velocity Measurement and Image Processing for Accurate Shape Restoration

Several steps were taken to obtain the length in the Y-axis direction of the particles. Firstly, the background images from the two images of the particle (captured by two different wavelength lights) were subtracted. Thereafter, these images are binarized by a threshold set in consideration of the gray-scale of the particles. As these binary images are likely to contain noise cracks, the morphological operation must be conducted to eliminate these imperfections. We counted the most extended Y-axis length of each particle and calculated the velocity based on the principle.

Next, the correct shapes of the particles were restored by utilizing the calculated velocity. By multiplying the speed and the light irradiation time, the extensions of the particles can be calculated. As the amount of the elongations are determined by this procedure, copying the gray-scale values of the parts that were subjected to elongation back to the original positions will restore the original shape and area of the flowing particles. The area was determined from the binary image, and a comparison was made before and after the processing. All of this was done by Python 3.8 (Python Software Foundation (PSF), Wilmington, DW, USA).

4. Results and Discussion

4.1. Velocity Evaluation with Simultaneous Acquisition of the Two Images with Different Exposure Time

To acquire two pictures of different wavelengths simultaneously, we fabricated the following set-up of the system. Figure 2a is a schematic illustration of the image acquisition flow cytometer with simultaneous two-wavelength differential image acquisition and analysis. The image acquisition interval was 5000 μs, and the two light sources were irradiated in synchronization. The irradiation time of the halogen lamp was set to 4994 μs by the camera's electronic shutter, and the LED was set to 2500 μs. These two lights were mixed at the dichroic mirror unit and followed the same optical path, and irradiated the sample in the microchip. The image of particles within the microchannel of the microchip was acquired by the 20x objective lens. After passing through the multi-view unit, the mixed image was separated into two pictures of different wavelengths of light and aligned to the CCD camera side-by-side after cutting each image to meet the 1/2 size of the CCD module area. By the above process, an image with two wavelengths of light can be acquired simultaneously in a single shot. The difference of their elongation can be compared side-by-side only within a picture of CCD image recording.

Figure 2b shows the optical path image diagram inside the multi-view unit. The light that was a mixture of two-wavelength was passed through the 2/3 frame window for cutting the acquired image to meet the size of 1/2 of the CCD module area, and was separated into two different wavelengths by the dichroic mirror A. These lights were controlled by the adjusted mirrors and were positioned to the left half or right half of the CCD camera's detector to be aligned side by side.

Figure 2c is an image of cells flowing through a microchannel in a microchip. In this experiment, we adopted hydrodynamic focusing to compare the distribution of particles before and after focusing. The samples were applied to the upper center stream and focused at the center of the pathway by the two side sheath flows. Then the samples were gathered to the center in the downstream area.

A schematic diagram of the microchip is described in Figure 2d. The sample and sheath buffer was placed in specified inlets, and the same air pressure was applied to these three inlets simultaneously to flow them into the microchannel equally.

Figure 2e shows the cross-sectional view of sample flow in the microchannel using air pressure. After the sample and the sheath buffers were inserted inside the specific inlets, a silicon cap lid was attached to seal the whole three inlets of the channels to apply the same air pressures for the balance of flow speeds. The samples and the sheath buffers were pushed by the applied air pressure and flow into the microchannels. The applied air pressure was controlled by the syringe pump and monitored. For example, when the applied pressure was 1.5 kPa, the max flow velocity at the center of the pathway was around 1 mm/s.

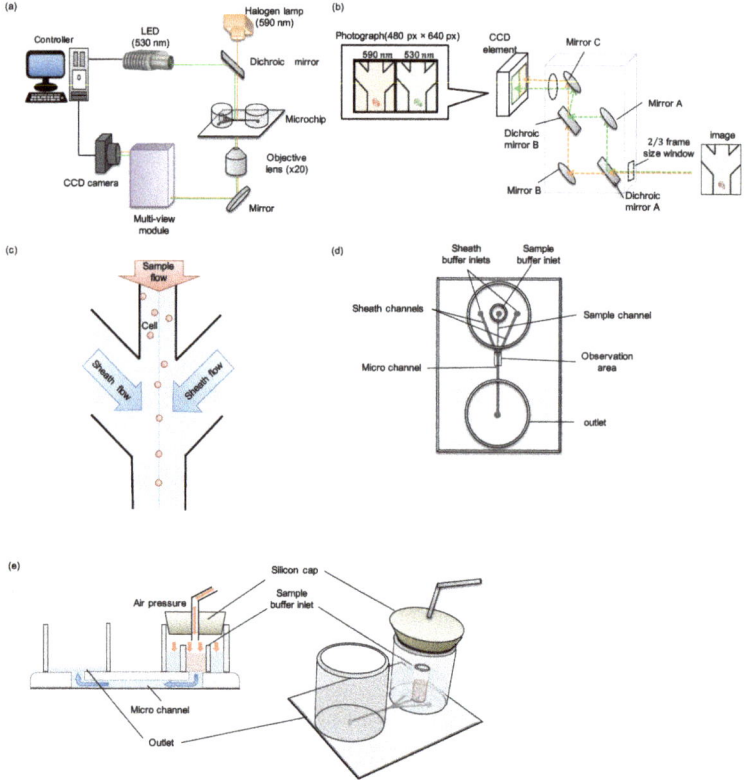

Figure 2. Set-up of the image acquisition flow cytometer with simultaneous two-wavelength differential imaging. (**a**) Equipment diagram of image acquisition flow cytometer system using simultaneous two-wavelength differential imaging. The system consists of nine modules. (i) Halogen lamp with band-pass filter (wavelength: 590 nm, irradiation time: 4994 μs); (ii) LED (wavelength: 530 nm, irradiation time: 2500 μs); (iii) Dichroic mirrors that reflect light below 550 nm—it was used for inserting LED in the light path of the halogen lamp; (iv) microchip; (v) objective lens (×20); (vi) mirror; (vii) multi-view unit; (viii) high-speed Charge-Coupled Device (CCD) camera (200 fps); (ix) controller. (**b**) Multi-view unit schematic diagram. The device adopts the mirror and the dichroic mirrors to enable the separation of an image into two at a wavelength of 550 nm. (**c**) An image of cells flowing through a flow cytometer. The cells converge in the flow channel with the sheath buffer. (**d**) Microchip schematic diagram. The microchip consists of polydimethylsiloxane (PDMS) on which the tubes are attached to the surface to create the inlets and the outlets. (**e**) The schematic diagram of the sample flows into the microchannel using air pressure.

First, we examined the principle of velocity measurement with fixed 12 μm polystyrene spheres under the dry condition on the mechanically controlled movement apparatus. A beam chopper (MC1000; THORLABS, Newton, NJ, USA) was inserted to the position of the microchip within the system and the movement of the particle fixed in the microchip on the beam chopper was observed.

For the velocity measurement, the images of samples were cut around the particle at 80 pixels (px) × 80 px from the full size of images of 530 and 590 nm wavelengths, and those images of the particles at two wavelengths with different irradiation times were saved as PNG files into the analysis computer. After the subtraction of the background with no prerecorded sample images, the 8-bit (256-step) grayscale values of obtained particle images were transformed into binary images using threshold values based on the average intensities of the acquired images of each wavelength. Using binarized images, the lengths of the X-axis, Y-axis, and the areas of particles were determined by counting the white pixels.

Figure 3a shows an example of a set of images of the single microparticle fixed in the channel. The upper two bright-field images are the micrographs of 590-nm with 4994 µs (left) and 530-nm with 2500 µs (left) (right) lights. These two images were acquired simultaneously as a single shot image in a CCD-camera. They were divided from the single-shot image of the particle with dichroic mirrors in the multi-view module. The lower two micrographs are their binarized images to measure their elongation automatically on a computer.

As shown in the images in Figure 3a (i), the acquired images of fixed particle shapes with different wavelengths and exposure time showed the same spherical shape of the single 12 µm polystyrene sphere. When we apply the 20x obj. lens with HXC13 CCD camera, the binarized data values of both wavelength images were the same and were 21 px in diameter for the X-axis (perpendicular to the flow direction) and 21 px in diameter for Y-axis (parallel to flow direction). The area of the particle was 360 px. As the size of one pixel was 0.704 µm, the diameter of the particles can be measured as 14.7 µm. The results confirmed no significant difference in image size and length caused by the difference in the wavelength of lights.

Although the beads were spherical during their fixed stationary position (stopping) (Figure 3a (i)), elongated particle images were acquired due to the movement of the particles during the light exposure time (Figure 3a (ii)). When the microbead moved to Y-direction (flow direction) with 1.4 mm/s mechanically by rotation of the beam chopper, the binarized images were changed to 19 (X-axis), 28 (Y-axis), and 472 px (area) for the left image (590 nm with 4994 µs), and 19 (X-axis), 23 (Y-axis), and 359 px (area) for the right image (530 nm with 2500 µs) (see Figure 3a (ii)). The difference of the length of particles caused by the different irradiation time of lights was used for estimating the velocity of particle movement and was $(28 - 23)/(4994 - 2500) \times 0.704 = 1.41 \times 10^{-3}$ [m/s], which was consistent with the rotation velocity. It should be noted that the X-axis length of both particles was similarly shrunk by 2 px, which should not be changed in this experiment because this direction is perpendicular to the movement direction. It may be caused by the image blur of the total shape of the particle and might have changed the edge of binarized images of particles. In this experiment, we did not add any edge enhancement technology into the edge detection and applied same threshold value for moving particles as the stopped particles. When we applied more precise edge detection technology, more precise binarized data analysis can be adopted for this velocity measurement.

As shown in the images of Figure 3a (ii), the difference of elongation ratio depended on the difference of irradiation time of two-wavelength lights. In other words, the exposure time changes the shapes of the sample in images. However, using the acquired velocity information, we can restore the original shape of the samples by exploiting the idea explained in Figure 1b (iv). The principle-based reconstruction was shown in Figure 3a (iii). The pixels of the lower half in the flow direction of the raw image of the sample was lifted to upstream direction by 10 px in one by one manner for the left image and by 5 px for the right image, where the values 10 px and 5 px were acquired from the calculation of the shift of the lower edge position pixels during the 4994 µs and 2500 µs exposure time with a flow velocity of particle, 2 px/ms.

Figure 3b demonstrates the result of the accuracy of the velocity measurement of this principle. The velocity of movement of the microparticle on the rotating beam chopper was observed at 0, 483.7, 725.4, 967.0, and 1208.3 mm/s, and the irradiation time of lights was set at 2500 and 4994 µs for 530 and 590 nm lights, respectively. Exploiting the velocity measurement procedure described above, we plotted the relationship between the set rotation velocity of the beam chopper and the measured velocity of the 12 µm polystyrene spheres. The results exhibited a linear correlation between the set velocities and measured velocities. Therefore, the result authenticated that the particle velocity can be measured with the single-shot of two-wavelength images using this method.

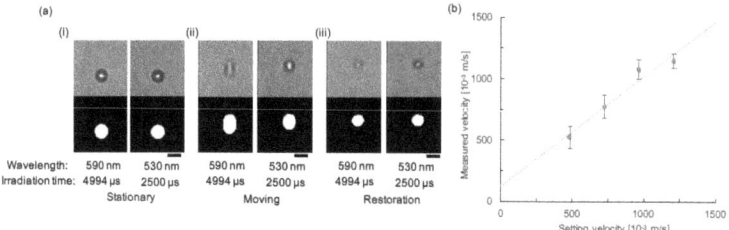

Figure 3. Results of principle experiment. (a) Original images (upper) and binarized images (lower) of the 12 μm single polystyrene sphere in the microchannel taken with an irradiation time of 4994 (left) and 2500 μs (right). (i) Images of stationary (stopping) particle. (ii) Images of moving particles. Elongation of the particle shapes was different depending on the difference of irradiation time. (iii) Restored images after image processing based on the calculated velocity with this method. Bars, 15 μm (1 px = 0.704 μm). (b) The relationship between the preset rotation velocity of the beam chopper (particle) and the measured velocity based on the principle (Plotted data and bars are mean values and standard deviations: N = 51, 53, 52, and 54 samples for 483.7, 725.4, 967.0, and 1208.3 mm/s of setting velocities, respectively). The dashed line indicates the linear fitting result ($y = 0.8982x + 122.62$, $R^2 = 0.952$).

4.2. Flow Velocity Distribution in Microfluidic Pathway

Next, we examined the ability of flow velocity measurement of microparticles within the microchannel with shingle-shot image acquisition. In this experiment, 75 μL of the sample was applied to the sample inlet of the microchip in the system. When the sample was microbeads, pure water was used as the buffer solution and sheath buffer, and when the sample was cells, the cell suspension was used as a sheath buffer. Air pressure was applied to both sample and sheath buffer inlets equally and simultaneously using a syringe pump to control the flow speed of samples (Figure 2e). The microchip was illuminated by a halogen lamp that emits light continuously and an LED light with a 2.5 ms flushing every 5 ms intervals synchronized with the shutter opening timing of the CCD camera.

Figure 4 shows the measurement results of the velocity of 12 μm microbeads flowing in the microchannel. Figure 4a is an image of the PDMS microchannel, indicating the two data acquisition positions: one (data acquisition position 1) was the upper stream of sample inlet pathway before the junction of hydrodynamic focusing, and the other (data acquisition position 2) was the lower stream after hydrodynamic focusing where the sheath side buffers were expected to concentrate the samples into the center of the microchannel. Particles that have passed through these positions were observed and stored as captured images into the computer.

Figure 4b shows the spatial distribution of passed 142 particles at the upper stream (data acquisition position 1). The distribution demonstrates that the flow of the particles was dispersed all over the microfluidic pathway width. Figure 4c shows the spatial distribution of the measured velocity of 142 particles in the microfluidic pathway (X position) at the upper stream (data acquisition position 1). The mean values and standard deviations of particles were plotted in the graph, and the dashed line was the curve fitting result. As described in the dashed line, the obtained results were consistent with the expected laminar flow in the microchannel. Figure 4d is the velocity distribution of 142 particles at the upper stream (data acquisition position 1). As the histogram indicates, a wide variety of flow velocity exists in the microchannel.

Figure 4e shows the spatial distribution of passing 91 particles at the lower stream (data acquisition position 2). It was confirmed that the particles were centered by the hydrodynamic focusing of two side sheath buffer flows. Figure 4f shows the spatial distribution of the measured velocity of 91 particles in the microfluidic pathway (X position) at the lower stream (data acquisition position 2). The mean values and standard deviations of particles were plotted in the graph, and the dashed line was the curve fitting result. Figure 4g shows the velocity distribution of 91 particles at the lower stream (data acquisition position 2). Although the spatial distribution of the particles was focused on the center of microchannel by the hydrodynamic focusing, the flow velocity distribution still remained. This result suggests that the spatial focusing was not sufficient to reduce the distribution of flow velocities of particles.

From the above results, we can indicate our system was able to measure the velocity of particles in the microchannel with only one shot. In addition, we found that the maximum flow velocity of microparticles differs depending on the X position in the channel, and even though the X position was the same, the flow velocity still varied even after the hydrodynamic focusing. Based on these results, we can suggest that it is necessary for accurate flow velocity measurement of each particle for precise cell sorting downstream because we need to estimate the correct sorting time to shift a particular target particle at the sorting point with the accurate flow velocity.

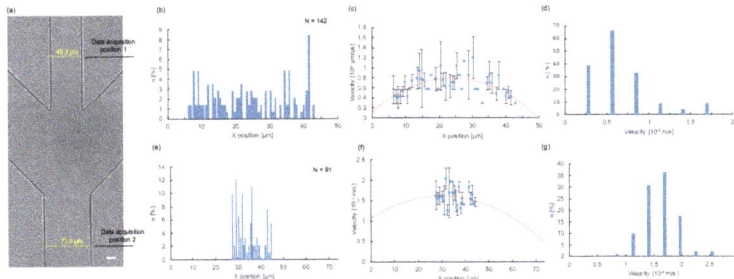

Figure 4. Measurement results of the velocity of microbeads flowing in the microchannel. (**a**) Microchannel and two data acquisition positions. The flow path widths for the upper stream (data acquisition positions 1) and the lower stream (data acquisition position 2) were 49.3 and 73.9 μm, respectively. The depth was 19.9 μm. Bar, 15 μm (1 px = 0.704 μm). (**b**) Positional distribution of 142 passing particles at the upper stream (data acquisition position 1). We counted the center coordinates of the particles. (**c**) Relationship between X position and measured velocity of 142 particles at the upper stream (data acquisition position 1). Plots show the average velocity and standard deviation of particles at each X position. The dashed line indicates the quadratic approximation curve ($y = -0.0012x^2 + 0.0567x + 0.1255, R^2 = 0.350$). (**d**) Velocity distribution of whole 142 particles at the upper stream (data acquisition position 1). (**e**) Positional distribution of 91 passing particles at the lower stream (data acquisition position 2). (**f**) Relationship between X position and measured velocity of 91 particles at the lower stream (data acquisition position 2). Plots show the average velocity and standard deviation of particles at each X position. The dashed line indicates the quadratic approximation curve ($y = -0.0006x^2 + 0.0364x + 1.0808, R^2 = 0.037$). (**g**) Velocity distribution of whole 91 particles at the lower stream (data acquisition position 2).

4.3. Simultaneous Flow Velocity Measurement of Particles in Microfluidic Pathway

Another advantage of this flow velocity measurement method is the simultaneous measurement of the plurality of particles with a single shot of image acquisition. We examined the ability of this simultaneous velocity measurement of particles.

Figure 5 shows an example of the velocity measurement of the plurality of particles in a shot of image acquisition flowing in the microchannel. Figure 5a shows a photograph of five particles flowing in the microchannel simultaneously within the image (left, 4994 μs exposure of 590 nm image, and right, 2500 μs of 530 nm image). Figure 5b is the relationship between the width (X) position and the velocity of observed five particles. From this result, the velocity distribution of microparticles were followed by the parabolic laminar flow distribution. As described above, we showed that our system can measure the velocities of all particles within a single shot of an image.

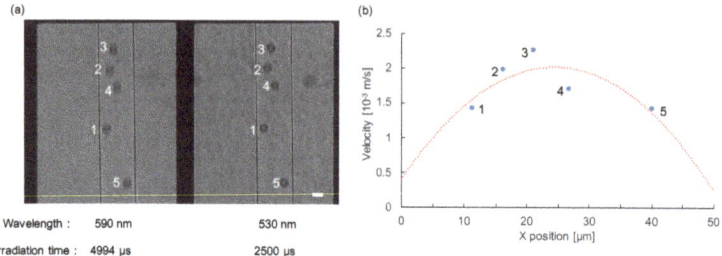

Figure 5. Velocity measurement of simultaneously flowing five particles. (**a**) The image was taken when multiple particles flowed simultaneously in the microchannel. Two-wavelength images of five particles were observed. The elongation was different depending on the difference of their flow velocities. Bar, 15 µm (1 px = 0.704 µm). (**b**) Measurement results of the five observed particle velocities. The velocity was dependent on the location of microchannel. The red dashed line is the quadratic approximation curve of laminar flow ($y = -0.0027x^2 + 0.133x + 0.3987$, $R^2 = 0.610$).

4.4. Precise Size Measurement with Flow Velocity Correction of Particles in Microfluidic Flow

Imaging flow cytometry uses the images of particles to identify target samples instead of their indirect information, such as the intensity of diffraction or fluorescence. Hence, the precise measurement of the image-based index, including actual size or a particular shape, is essential. However, because of the movement of samples during the exposure time, the acquired images of samples were elongated to the flow direction as far as the exposure time exists. To overcome this problem, our single-shot velocity measurement method can also give us another advantage for improving the accuracy of sample sizes. When we can acquire the precise flow velocity of each particle, we can reconstruct the shape information of the particle based on this displacement information—as explained in Figure 1b (iv).

Figure 6 shows the result of the actual shape restoration process of the same 142 samples of Figure 4b–d. Figure 6a shows the images of a stationary (stopping) particle in 590 and 530 nm wavelengths (upper) and their binarized images (lower). These binarized images of the 12 µm microbead in 530 and 590 nm wavelengths were round-shaped. Figure 6b shows images of a moving (elongated) particle and its binarized pictures in 530 and 590 nm wavelengths. The difference in elongation lengths was caused by the difference in the exposure time, 4994 and 2500 µs. From the difference in particle's elongation length and the difference in irradiation time of these two light sources, the velocity of this particle was determined as 1.69×10^{-3} m/s. Figure 6c shows images obtained by the conventional restoration method, which just compressed the raw image of the flow direction (Y-axis) to recover the amount of elongation that occurred during the exposure time. The irradiation time of the light having a wavelength of 590 nm is 4994 µs. By using the irradiation time of the light and the calculated velocity (1.69×10^{-3} m/s), the particle's elongation was identified to be 8.44 µm. Since the length of the elongated particles was 14.8 µm, As shown in Figure 6c, the elongated shape of the particle was recovered to 12 µm in the flow direction, which was obtained with the magnification of the elongated particle image at 8.44/14.8 in the flow direction. However, the shape of the particle was different from the spherical microbead. Hence, we can conclude that the conventional simple image reduction method fails to restore the original shape of the particles.

Figure 6d is the image obtained from the restoration method we proposed in Figure 1b (iv), in which the lower parts of the elongated particle were lifted by the amount of elongation determined by multiplying the calculated velocity and irradiation time. As shown in the images, the original shape and size of the circular particle have been restored appropriately. Figure 6e shows the relationship between the X position and the averaged area of 142 samples of Figure 4b–d. In general, as shown in Figure 6e (i), even though all the particle sizes were 12 µm, the acquired raw area data of those samples were larger when the velocity of the particles was faster. When we applied the conventional method to all elongated particle images, the distribution of the difference in their areas was shown depending on the X position, and it can be seen that this method cannot restore the correct shape (Figure 6e (ii)). On the other hand, when we applied our restoration method to the particles, the distribution of particle areas became flat (Figure 6e (iii)). These results indicate that the importance of the correct restoration method to acquire accurate area information and also the importance of the acquisition of precise velocity information of each particle for its restoration.

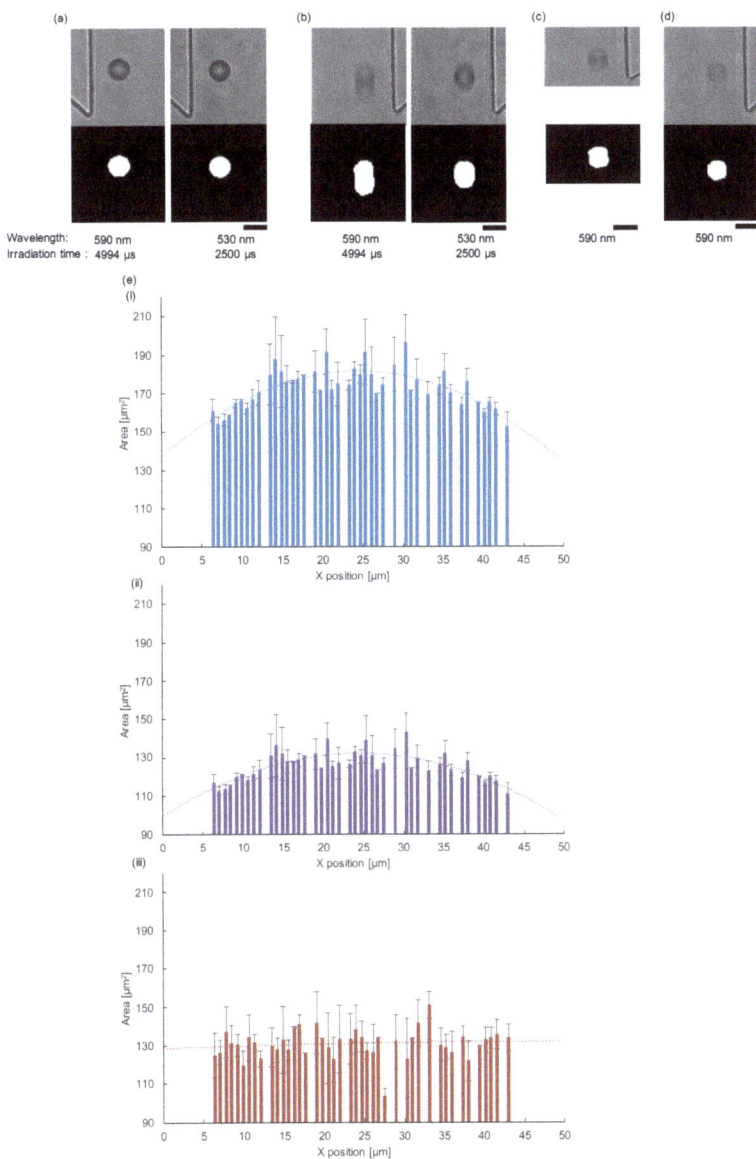

Figure 6. Result of exact shape restoration process considering the particle's velocity. (**a**) Images of a stationary (stopping) particle (upper) and its binarized images (lower). (**b**) Images of flowing particles. Bright-field images (upper) and binarized images (lower). (**c**) Images obtained from conventional

restoration method, which compresses the elongated particle image only considering the flow velocity. The elongated shape was equally shortened in the flow direction. (**d**) Images obtained from the restoration processing considering velocity and exposure time, as explained in Figure 1b (iv). The acquired shape of the particle was recovered to the round shape and was similar to the stationary (stopping) particle. Bars, 15 µm (1 px = 0.704 µm). (**e**) Relationship between the position in the microchannel (X-axis) position and the area of 142 microbeads shown in Figure 4b–d. (i) The raw image data of particle area distribution. The areas of particles were correlated to the flow velocities of those particles. Hence, the raw data of particle area distribution was well fit to the parabolic curvature (dashed line: $y = -0.0744x^2 + 3.6406x + 137.26$, $R^2 = 0.615$). (ii) The corrected particle area distribution only considering the flow velocity. Dashed line was the averaged area of particles acquired and still shows a quadratic curve ($y = -0.0541x^2 + 2.6477x + 99.826$, $R^2 = 0.615$). (iii) The corrected particle area distribution considering velocity and exposure time. The dashed line was the mean value of the corrected area ($-0.0015x^2 + 0.1404x + 128.64$, $R^2 = 0.010$) and was almost flat regardless of the place in the microchannel.

4.5. Ability and Potential Limitation of This Method for Imaging Flow Cytometry Measurement

In this paper, we proposed the simple single-shot image-based velocity measurement method and its application for size correction. This method exploited the exposure time for image acquisition for velocity analysis. Usually, the exposure time is necessary to the image acquisition and the origin of quality reduction of images, namely image blur. However, focusing on the bright side of these characteristics enables us to acquire the precise measurement of velocity and size of samples as described above in this paper.

In principle, this method is simple and can be applied for all the measurements. However, the resolution and preciseness are limited to the ability and resolution of hardware. From this viewpoint, the ability of this method is strongly reliant on the progress of the technologies.

First, we focused on the limitations of camera resolution. If the moving distance of the particle is less than the minimum unit of resolution (1 px) during the light irradiation time, the image of the particle is obtained without stretching. Figure 7a shows the relationship between particle velocity and maximum irradiation time. Capturing the particles without elongation is possible by setting the irradiation time below the time of the curve shown in the figure. On the other hand, since the velocity is calculated using the difference in the particle's Y-axis lengths, the difference between the irradiation times of the two wavelengths needs to be longer than the maximum irradiation time.

Next, we considered the influence on sorting accuracy. If multiple particles appear in the image clipped to 80 × 80 px, accurate particle recognition cannot be performed. Therefore, it is necessary to adjust the density and velocity of the sample particles. Figure 7b shows the relationship between the speed and maximum density of 12 µm beads at the data acquisition position 2. The minimum distance that prevents neighboring particles from being simultaneously captured was calculated and converted to density. For this purpose, the elongation of the particle, determined from the velocity and the irradiation time, was taken into account. Shutter time was set to 4994 µs. A depth of 19.9 µm, a mean X position of the flowing particles of 34.4 µm, and a standard deviation of 5.31 µm at data acquisition position 2 (Figure 4f) were used in the calculation. In order to perform an accurate measurement, it is necessary to measure the sample at a density lower than the max density indicated by the straight line in the figure.

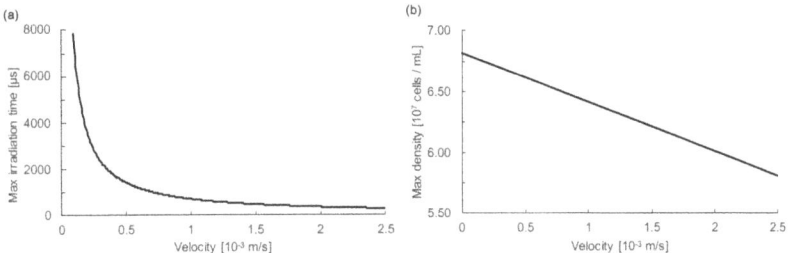

Figure 7. The potential limitation of this method. (**a**) Relationship between particle velocity and maximum irradiation time. The maximum irradiation time is the time required for a particle to move 1 px (1 px = 0.704 μm). If the irradiation time is shorter than that time, it is possible to obtain an image of the particles without stretching. (**b**) Relationship between particle velocity and the maximum density of 12 μm beads. This refers to a density that does not allow multiple particles to appear in the 80 × 80 px image that is cropped. The irradiation time was set to 4994 μm, and the elongation of the particles due to this was taken into account.

4.6. Accurate Shape Reconstruction of Flowing HeLa Cells

Finally, we applied this method for practical living samples by using HeLa cells. Figure 8 shows images of a flowing HeLa cell at the data acquisition position two (see Figure 4a) and its restored images in 590 and 530 nm wavelengths. The acquired cell's velocity was 1.41×10^{-3} m/s. The area of the cell in the captured image (590 nm) was 293.5 μm², whereas its size was corrected to 170.6 μm² after restoration. This example shows the ability of this method for the living cells.

Recently the reconstruction technology of 3-D images of cells and their clusters by their rotations in the microfluidic device was developed [18]. Our precise cell flow velocity can give precise displacement information adding to the rotation images of particles to reconstruct those 3-D images even during flowing.

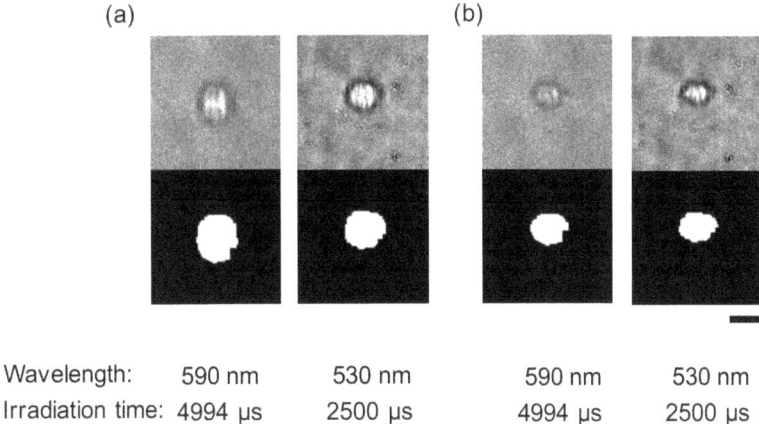

Figure 8. Results of a restoration of a flowing HeLa cell. (**a**) Images of a flowing HeLa cell (upper) and its binarized images (lower) at data acquisition position 2. Bar, 15 μm (1 px = 0.704 μm). (**b**) Images obtained from the restoration processing considering velocity and exposure time.

5. Conclusions

We examined our simple single-shot measurement of the flow velocity of samples and its application for sample size restoration in standard 12 μm polystyrene spheres and HeLa cells. This method exploited the positive side of exposure time of images and can obtain all the sample velocities and corrected area size information simultaneously with a single image acquisition shot. Although the resolution and preciseness of acquired velocity and area information were limited by the ability and resolution of existing hardware, in principle, this method can be applied for all the imaging flow cytometry, and also the ability and resolution can be improved according to the improvement of hardware. Especially the advantage of this method, precise measurement of the flow velocity of single particles within a flow, can be applied for correction and reconstruction of images such as high-throughput reconstruction of 3-D images of flowing cell clusters by their rotations as the next step of this application. We think this method can contribute to the improvement of general imaging flow cytometry for more precise target recognition and also for correct target collection timing decisions especially for the diagnostic screening including circulation tumor cells.

Author Contributions: Conceptualization, K.Y.; methodology, K.Y. and S.S.; software, S.S.; analysis and validation, S.S., M.S., and K.Y.; writing, S.S., M.S., and K.Y.; supervision, K.Y.; funding acquisition, K.Y. All authors have read and agreed to the published version of the manuscript.

Funding: This research was funded by research and development projects of the Industrial Science and Technology Program, the New Energy and Industrial Technology Development Organization (NEDO, P08030), JSPS KAKENHI Grant Number JP17H02757, JST CREST program, and Waseda University Grant for Special Research Projects (2016S-093, 2017B-205, 2017K-239, 2018K-265, 2018B-186, 2019C-559), the Leading Graduate Program in Science and Engineering for Waseda University from MEXT, Japan.

Acknowledgments: We would like to thank Akihiro Hattori, Masao Odaka, and all the members of Yasuda lab for their technical supports, discussion, and suggestions.

Conflicts of Interest: The authors declare no conflict of interest.

Abbreviations

The following abbreviations are used in this manuscript:

CAD	computer-aided design
CTC	circulating tumor cell
DMEM	Dulbecco's modified Eagle medium
FBS	fetal bovine serum
LED	light emitting diode
PDMS	polydimethylsiloxane
px	pixel

References

1. Chou, H.P.; Spence, C.; Scherer, A.; Quake, S. A microfabricated device for sizing and sorting DNA molecules. *Proc. Natl. Acad. Sci. USA* **1999**, *96*, 11–13, doi:10.1073/pnas.96.1.11.
2. Cheung, K.; Gawad, S.; Renaud, P. Impedance spectroscopy flow cytometry: On-chip label-free cell differentiation. *Cytom. Part A* **2005**, *65*, 124–132, doi:10.1002/cyto.a.20141.
3. Huh, D.; Gu, W.; Kamotani, Y.; Grotberg, J.B.; Takayama, S. Microfluidics for flow cytometric analysis of cells and particles. *Physiol. Meas.* **2005**, *26*, R73–R98, doi:10.1088/0967-3334/26/3/R02.
4. Cheung, K.C.; Berardino, M.D.; Schade-Kampmann, G.; Hebeisen, M.; Pierzchalski, A.; Bocsi, J.; Mittag, A.; Tárnok, A. Microfluidic impedance-based flow cytometry. *Cytom. Part A* **2010**, *77*, 648–666, doi:10.1002/cyto.a.20910.
5. Bow, H.; Pivkin, I.V.; Diez-Silva, M.; Goldfless, S.J.; Dao, M.; Niles, J.C.; Suresh, S.; Han, J. A microfabricated deformability-based flow cytometer with application to malaria. *Lab Chip* **2011**, *11*, 1065–1073, doi:10.1039/c0lc00472c.

6. Karabacak, N.M.; Spuhler, P.S.; Fachin, F.; Lim, E.J.; Pai, V.; Ozkumur, E.; Martel, J.M.; Kojic, N.; Smith, K.; Chen, P.I.; et al. Microfluidic, marker-free isolation of circulating tumor cells from blood samples. *Nat. Proc.* **2014**, *9*, 694–710, doi:10.1038/nprot.2014.044.
7. Stott, S.L.; Hsu, C.H.; Tsukrov, D.I.; Yu, M.; Miyamoto, D.T.; Waltman, B.A.; Rothenberg, S.M.; Shah, A.M.; Smas, M.E.; Korir, G.K.; et al. Isolation of circulating tumor cells using a microvortex-generating herringbone-chip. *Proc. Natl. Acad. Sci. USA* **2010**, *107*, 18392–18397, doi:10.1073/pnas.1012539107.
8. Hosokawa, M.; Yoshikawa, T.; Negishi, R.; Yoshino, T.; Koh, Y.; Kenmotsu, H.; Naito, T.; Takahashi, T.; Yamamoto, N.; Kikuhara, Y.; et al. Microcavity array system for size-based enrichment of circulating tumor cells from the blood of patients with small-cell lung cancer. *Anal. Chem.* **2013**, *85*, 5692–5698, doi:10.1021/ac400167x.
9. Yu, B.Y.; Elbuken, C.; Shen, C.; Huissoon, J.P.; Ren, C.L. An integrated microfluidic device for the sorting of yeast cells using image processing. *Sci. Rep.* **2018**, *8*, 1–12, doi:10.1038/s41598-018-21833-9.
10. Utharala, R.; Tseng, Q.; Furlong, E.E.; Merten, C.A. A Versatile, Low-Cost, Multiway Microfluidic Sorter for Droplets, Cells, and Embryos. *Anal. Chem.* **2018**, *90*, 5982–5988, doi:10.1021/acs.analchem.7b04689.
11. Sesen, M.; Whyte, G. Image-Based Single Cell Sorting Automation in Droplet Microfluidics. *Sci. Rep.* **2020**, *10*, 1–14, doi:10.1038/s41598-020-65483-2.
12. Nitta, N.; Sugimura, T.; Isozaki, A.; Mikami, H.; Hiraki, K.; Sakuma, S.; Iino, T.; Arai, F.; Endo, T.; Fujiwaki, Y.; et al. Intelligent Image-Activated Cell Sorting. *Cell* **2018**, *175*, 266–276.e13, doi:10.1016/j.cell.2018.08.028.
13. Ota, S.; Horisaki, R.; Kawamura, Y.; Ugawa, M.; Sato, I.; Hashimoto, K.; Kamesawa, R.; Setoyama, K.; Yamaguchi, S.; Fujiu, K.; et al. Ghost cytometry. *Science* **2018**, *360*, 1246–1251, doi:10.1126/science.aan0096.
14. Andree, K.C.; van Dalum, G.; Terstappen, L.W. Challenges in circulating tumor cell detection by the CellSearch system. *Mol. Oncol.* **2016**, *10*, 395–407, doi:10.1016/j.molonc.2015.12.002.
15. Hao, S.J.; Wan, Y.; Xia, Y.Q.; Zou, X.; Zheng, S.Y. Size-based separation methods of circulating tumor cells. *Adv. Drug Deliv. Rev.* **2018**, *125*, 3–20, doi:10.1016/j.addr.2018.01.002.
16. Huang, X.; Tang, J.; Hu, L.; Bian, R.; Liu, M.; Cao, W.; Zhang, H. Arrayed microfluidic chip for detection of circulating tumor cells and evaluation of drug potency. *Anal. Biochem.* **2019**, *564–565*, 64–71, doi:10.1016/j.ab.2018.10.011.
17. Shibuta, M.; Tamura, M.; Kanie, K.; Yanagisawa, M.; Matsui, H.; Satoh, T.; Takagi, T.; Kanamori, T.; Sugiura, S.; Kato, R. Imaging cell picker: A morphology-based automated cell separation system on a photodegradable hydrogel culture platform. *J. Biosci. Bioeng.* **2018**, *126*, 653–660, doi:10.1016/j.jbiosc.2018.05.004.
18. Puttaswamy, S.V.; Bhalla, N.; Kelsey, C.; Lubarsky, G.; Lee, C.; McLaughlin, J. Independent and grouped 3D cell rotation in a microfluidic device for bioimaging applications. *Biosens. Bioelectron.* **2020**, *170*, 112661, doi:10.1016/j.bios.2020.112661.
19. Takahashi, K.; Hattori, A.; Suzuki, I.; Ichiki, T.; Yasuda, K. Non-destructive on-chip cell sorting system with real-time microscopic image processing. *J. Nanobiotechnol.* **2004**, *2*, 5, doi:10.1186/1477-3155-2-5.
20. Hayashi, M.; Hattori, A.; Kim, H.; Terazono, H.; Kaneko, T.; Yasuda, K. Fully automated on-chip imaging flow cytometry system with disposable contamination-free plastic re-cultivation chip. *Int. J. Mol. Sci.* **2011**, *12*, 3618–3634, doi:10.3390/ijms12063618.
21. Yasuda, K.; Hattori, A.; Kim, H.; Terazono, H.; Hayashi, M.; Takei, H.; Kaneko, T.; Nomura, F. Non-destructive on-chip imaging flow cell-sorting system for on-chip cellomics. *Microfluid. Nanofluid.* **2013**, *14*, 907–931, doi:10.1007/s10404-012-1112-6.
22. Girault, M.; Kim, H.; Arakawa, H.; Matsuura, K.; Odaka, M.; Hattori, A.; Terazono, H.; Yasuda, K. An on-chip imaging droplet-sorting system: A real-time shape recognition method to screen target cells in droplets with single cell resolution. *Sci. Rep.* **2017**, *7*, 40072, doi:10.1038/srep40072.
23. Kim, H.; Terazono, H.; Nakamura, Y.; Sakai, K.; Hattori, A.; Odaka, M.; Girault, M.; Arao, T.; Nishio, K.; Miyagi, Y.; et al. Development of on-chip multi-imaging flow cytometry for identification of imaging biomarkers of clustered circulating tumor cells. *PLoS ONE* **2014**, *9*, e104372, doi:10.1371/journal.pone.0104372.

24. Hattori, A.; Kim, H.; Terazono, H.; Odaka, M.; Girault, M.; Matsuura, K.; Yasuda, K. Identification of cells using morphological information of bright field/fluorescent multi-imaging flow cytometer images. *Jpn. J. Appl. Phys.* **2014**, *53*.
25. Odaka, M.; Kim, H.; Nakamura, Y.; Hattori, A.; Matsuura, K.; Iwamura, M.; Miyagi, Y.; Yasuda, K. Size distribution analysis with on-chip multi-imaging cell sorter for unlabeled identification of circulating tumor cells in blood. *Micromachines* **2019**, *10*, 154, doi:10.3390/mi10020154.

Publisher's Note: MDPI stays neutral with regard to jurisdictional claims in published maps and institutional affiliations.

© 2020 by the authors. Licensee MDPI, Basel, Switzerland. This article is an open access article distributed under the terms and conditions of the Creative Commons Attribution (CC BY) license (http://creativecommons.org/licenses/by/4.0/).

Article

Scalable Parallel Manipulation of Single Cells Using Micronozzle Array Integrated with Bidirectional Electrokinetic Pumps

Moeto Nagai [1,*], Keita Kato [1], Satoshi Soga [1], Tuhin Subhra Santra [2] and Takayuki Shibata [1]

[1] Department of Mechanical Engineering, Toyohashi University of Technology, Toyohashi, Aichi 441-8580, Japan; bari91_ape50@yahoo.co.jp (K.K.); ss3104u5@gmail.com (S.S.); shibata@me.tut.ac.jp (T.S.)
[2] Department of Engineering Design, Indian Institute of Technology Madras, Tamil Nadu 600036, India; santra.tuhin@gmail.com
* Correspondence: nagai@me.tut.ac.jp; Tel.: +81-532-44-6701

Received: 21 March 2020; Accepted: 22 April 2020; Published: 22 April 2020

Abstract: High throughput reconstruction of in vivo cellular environments allows for efficient investigation of cellular functions. If one-side-open multi-channel microdevices are integrated with micropumps, the devices will achieve higher throughput in the manipulation of single cells while maintaining flexibility and open accessibility. This paper reports on the integration of a polydimethylsiloxane (PDMS) micronozzle array and bidirectional electrokinetic pumps driven by DC-biased AC voltages. Pt/Ti and indium tin oxide (ITO) electrodes were used to study the effect of DC bias and peak-to-peak voltage and electrodes in a low conductivity isotonic solution. The flow was bidirectionally controlled by changing the DC bias. A pump integrated with a micronozzle array was used to transport single HeLa cells into nozzle holes. The application of DC-biased AC voltage (100 kHz, 10 V_{pp}, and V_{DC}: −4 V) provided a sufficient electroosmotic flow outside the nozzle array. This integration method of nozzle and pumps is anticipated to be a standard integration method. The operating conditions of DC-biased AC electrokinetic pumps in a biological buffer was clarified and found useful for cell manipulation.

Keywords: micronozzle-array; parallel cell manipulation; bidirectional electrokinetic pump; DC biased AC electrokinetic flow

1. Introduction

Functional analysis of cells leads to new insights in the medical and pharmaceutical fields. Mimicking in vivo cellular environments in vitro is key to efficient analysis. In organs, cells communicate with each other via direct contact over short or long distances and are influenced by the cellular microenvironment, e.g., cell–cell and cell–matrix interactions. Heterotypic cell–cell interactions have been studied using in vitro co-culture systems [1–3]. Microfabrication and microfluidic technologies were used for co-cultures of cells and cell-cell interactions were studied [4,5]. However, conventional in vitro techniques are often insufficient for reconstructing cellular microenvironments in any combination. Single cell manipulation tools [6] are required for mimicking the in vivo environment in vitro with a higher reproducibility.

Single cells have been manipulated by fluidic [7–11], dielectrophoretic [12–14], and optical [15–17] techniques and have been encapsulated in droplets [18,19]. Previous methods had limited accessibility and did not offer sufficient versatility and flexibility. Accessibility to an open-top chip and high flexibility are provided by one-side-open devices such as a glass capillary [20,21], a cantilever with an aperture [22,23], dielectrophoretic tweezers [24,25], inkjet-like printing [26,27], and printed droplet microfluidics [28]. These devices are typically composed of a single probe, and their throughput is limited because of the limit on the number of channels. Increasing the number of channels enables higher throughput in manipulation of single cells while maintaining flexibility and open accessibility. The number of micro-channels for dispensing droplets containing single cells was increased to three, and three pneumatic pressure sources and solenoid valves were integrated into the system [29]. The manipulation of single cells was demonstrated with a hollow probe array and a single pressure source in open space [30], even though individual flow control of each micro-channel remained an issue.

Mechanical and electrical pumps [31] can be integrated with one-side-open multi-channel devices for individual flow control. Among these pumps, electroosmotic pumps (EOPs) [32] are easy to integrate with microdevices and scalable. DCEOPs have the advantage of simplicity [33,34] and bidirectional flow, but have a potential issue with bubble formation due to electrolysis. To avoid electrolytic reactions, ACEOPs have been developed. Although ACEOPs can change flow direction using a travelling wave [35,36], this method requires phase shift and is relatively difficult to scale. A DC-biased AC-electrokinetic (ACEK) flow was used to induce bidirectional flow [37]. W.Y. Ng et al. investigated the mechanism of the flow and concluded that a conductivity gradient drives fluid flow [38,39]. Deionized (DI) water [37,40], a 10^{-4} M solution of KCl (conductivity 1 mS m^{-1}), and a $CuSO_4$ solution (conductivity 4 mS m^{-1}) [38] were used as working fluids. However, the availability of an isotonic solution for DC-biased ACEKPs is unclear.

A method of integrating polydimethylsiloxane (PDMS) micro-nozzles, micro-channels, and pump electrodes for parallel manipulation of single cells was developed in this study. Interdigital tooth electrodes were made from indium tin oxide (ITO) or Pt/Ti, and a DC-biased AC signal applied to them. We characterized electrokinetic flow in an isotonic solution at both electrodes. The DC bias voltage was switched to control the direction of flow. Single cells were manipulated with a micro-nozzle array integrated with DC-biased ACEKPs. We characterized the manipulation of single cells with the nozzles.

2. Experimental Materials and Methods

2.1. Concept and Design of PDMS Nozzle Array Integrated with Electrokinetic Micropumps

The integration of DC-biased ACEK pumps into nozzles enables them to transport fluid in both directions (Figure 1). Since symmetric electrodes were used, the positive and negative of the DC bias were changed for bidirectional flow control. The diameter of micronozzles was designed to be around 30 µm for manipulating multiple types of single cells approximately 10–20 µm in diameter simultaneously. Cells are sucked into microchannels by the actuation of pumps (Figure 1a). The cells are trapped, moved to the top of microwells, and ejected from the channel to make a pair of cells (Figure 1b). An array of 4 × 4 micronozzles is integrated with DC-biased ACEKPs (Figure 1c), which are composed of seven pairs of electrodes.

The principles of the pump is based on DC-biased AC-electrokinetics [38]. Planar parallel electrodes are placed in a microfluidic channel in contact with an electrolyte solution and a DC biased AC electrical signal is applied to the electrode pair. The application of DC bias induces Faradaic electrolytic reactions and results in an increase of the ionic content of the bulk solution. The ionic contents differ at the cathodic and anodic sides. DC biased AC electric signal acts on the transverse conductivity gradient and generates fluid flow.

An array of 4 × 4 nozzles was arranged symmetrically with a pitch of 200 µm and concentrated in the middle region (680 µm × 680 µm) for microscopic observation of as many cells as possible. Each 60-µm width channel was connected to a 30-µm diameter nozzle for the concentration and the minimum spacing between channels was set at 30 µm. We followed the dimensions of electrodes for bidirectional flow transport from a previous study [37], where they were 80-µm wide with gaps of 20 µm and 100 µm. A lower-height channel induces higher flow velocity and we selected 70 µm as the channel height.

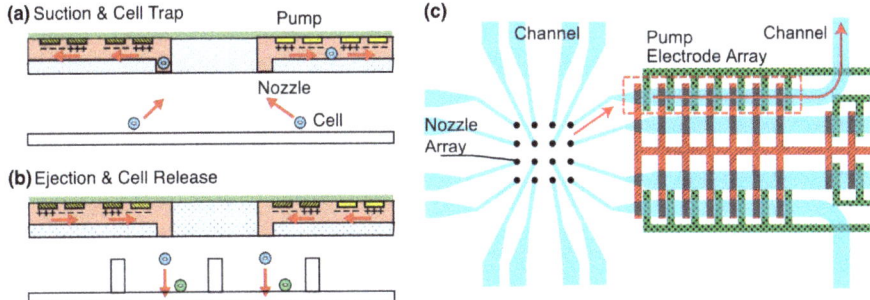

Figure 1. The schematic of a nozzle array integrated with electrokinetic pumps for cell manipulation. (a) Suction and trap of single cells in an array format with nozzles. (b) Ejection and release of the cells from the nozzles. (c) Schematic view of 4 × 4 nozzles integrated with pumps. One of the electrodes (green) is connected to a signal voltage (DC-biased AC voltage) and the other electrode (red) is connected to the ground.

2.2. Fabrication of PDMS Nozzle Array Integrated with Electrokinetic Micropumps

Then, 4 × 4 PDMS nozzles were integrated with electrokinetic ITO pumps for single cell manipulation. PDMS micronozzles were formed and combined with an ITO substrate patterned on a glass. First, a method of making PDMS though holes [41,42] for fabricating PDMS nozzles was used and the method extended from a single-layer mold to a double-layer mold. Double-layer thick photoresist, SU-8 3050 (Kayaku Microchem, Tokyo, Japan), was formed on a 3-inch Si substrate (Figure 2a). A silicon wafer was dehydrated at 160 °C for 5 min. The first layer of SU-8 was spincoated at 500 rpm for 25 s and 1000 rpm for 55 s to obtain a film thickness of 70 μm. It was pre-baked at 60 °C for 5 min, at 95 °C for 50 min, and at 60 °C for 5 min. The pattern was exposed through a PMMA filter and mask aligner (PEM-800, Union Optical Co., Ltd., Tokyo, Japan). The integrated exposure light was 4675 mJ/cm^2 (contact mode). The wafer was post-baked at 65 °C for 9 min, at 95 °C for 5 min, and at 65 °C for 2 min. The second SU-8 layer was spin-coated at 500 rpm for 25 s and 1000 rpm for 55 s to make the micronozzles after PDMS molding. The wafer was prebaked at 60 °C for 5 min, at 95 °C for 50 min, and at 60 °C for 5 min. The SU-8 was exposed to the integrated exposure light (4675 mJ/cm^2) through a mask and post-baked at 65 °C for 9 min, 95 °C for 5 min, and 65 °C for 2 min. The SU-8 molds of the microchannel or micronozzles were developed in 2-acetoxy-1-methoxypropane (Wako Pure Chemical Industries Ltd., Osaka, Japan) for 15 min. They were rinsed with isopropyl alcohol. The SU-8 molds were treated with vapor of trichloro(1H,1H,2H,2H-perfluorooctyl)silane (PFOCTS).

PDMS (Silpot 184, Dow Corning Toray Co., Ltd., Tokyo, Japan) was mixed at a 10:1 ratio of the base polymer to curing agent. Uncured PDMS was poured over the fabricated SU-8 molds (Figure 2b). A 3-inch diameter and 3-mm thick PDMS sheet was formed as the carrier sheet and treated with PFOCTS. Uncured PDMS was pressed with the PDMS carrier sheet and a weight of about 600 g (Figure 2c). PDMS was cured at room temperature for 24 h to expel PDMS from SU-8 pillars. The PDMS were released from the mold through holes on the sheet (Figure 2d).

A 200-nm thick ITO-coated 100 mm × 100 mm glass substrate (thickness 0.7 mm) was purchased from Geomatec Co., Ltd. (Yokohama, Japan) and used as the electrode (Figure 2e). A substrate was cut into a 33 mm × 33 mm piece and ultrasonically cleaned for 5 min with acetone, isopropyl alcohol (IPA), and DI water each. An ITO substrate was dehydrated at 160 °C for 5 min. Hexamethyldisilazane (HMDS) and OFPR-8600 (52 cp, Tokyo-Oka Kogyo Co., Ltd., Kawasaki, Japan) were spin-coated on the substrate at 1000 rpm for 5 s and 2000 rpm for 25 s. The thickness of OFPR was increased to resist etching in hydrochloric acid. The substrate was prebaked at 110 °C for 90 s. OFPR was exposed to an integrated light quantity of 100 mJ/cm^2 to form pump electrodes. The resistance was immersed in a developer (NMD-3, Wako Pure Chemical Industries, Ltd.) for 120 s, rinsed with DI water for 2 min, and blown with N_2 gas (Figure 2g). The resistance was post-baked at 140 °C for 5 min. ITO was etched in 9 mol/L hydrochloric acid at 40 °C for 6 min using OFPR as a mask (Figure 2h). The resistance on the bare ITO was checked to fully etch ITO. The resistance was removed by ultrasonic cleaning with acetone for 5 min. A 1-mm hole was drilled in the glass part of the ITO substrate for liquid introduction (Figure 2i).

The PDMS chip and ITO electrodes were treated with air plasma (plasma treater, YHS-R, Sakigake Semiconductor Co., Ltd., Kyoto, Japan) for 60 s for bonding. The substrates were aligned with mechanical stages and permanently bonded by baking them at 80 °C for 40 min (Figure 2j). The cured PDMS was peeled off from the PFOCTS-treated PDMS sheet. A PDMS sheet with 0.5-mm holes was bonded on the back side to introduce into the solution after plasma treatment. The micronozzles were integrated with the pumps (Figure 2k).

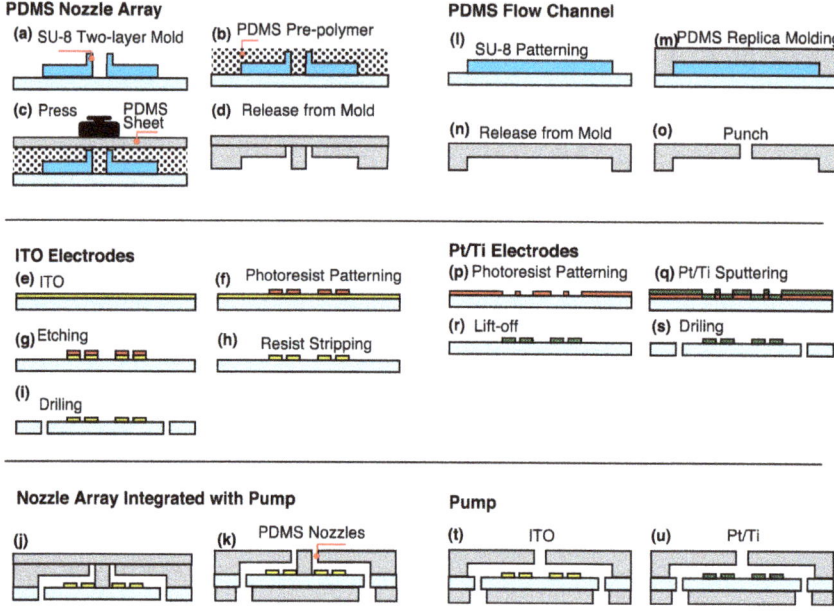

Figure 2. Fabrication process of the nozzle array integrated with pump. (**a–d**) Fabrication of polydimethylsiloxane (PDMS) nozzles. Uncured PDMS was pressed by a PDMS sheet to form PDMS through holes. (**e–i**) Patterning of OFPR photoresist on indium tin oxide (ITO) substrate and formation of ITO electrodes. ITO was etched with hydrochloric acid. (**j,k**) Bonding of the PDMS micronozzles, ITO electrodes, and a PDMS sheet after plasma treatment. (**l–o**) Fabrication of PDMS flow channel. (**p–s**) Patterning of Pt/Ti electrodes. (**t–u**) Bonding of the PDMS microchannel, ITO, or Pt/Ti electrodes, and a PDMS sheet after plasma treatment.

2.3. Fabrication of EOF Pump from ITO and Pt/Ti Electrodes

Interdigital electrodes were patterned on a glass substrate (Figure 2e–i,p–s) and bonded to a PDMS flow channel. A PDMS channel was molded from the SU-8 structure. The electrode was an interdigital structure in which seven symmetrical electrodes were arranged at regular intervals. ITO and Pt/Ti were used as the electrode material. For the Pt/Ti electrode, Ti was deposited as an adhesion layer of Pt and glass, and Pt was utilized as the electrode surface.

The PDMS flow channel was prepared as in Section 2.1. A single-layer thick photoresist, SU-8 3050, was used for a mold of the PDMS micro-channel (Figure 2l). PDMS was cured at 80 °C and microchannels were molded in PDMS from the SU-8 mold (Figure 2m) and released (Figure 2n).

A 0.5 mm-hole was punched at the end of the PDMS flow channel to introduce and eject a solution (Figure 2o).

A slide glass (76 × 26 × 1 mm, No. 1, Matsunami Glass Industry Co., Ltd., Osaka, Japan) was cut into three equal parts (about 26 mm × 26 mm) for the Pt/Ti electrodes. The glass was cleaned in a mixture of sulfuric acid and hydrogen peroxide mixture (SPM) and dehydrated at 140 °C (5 min). HMDS and OFPR-8600 were spin-coated at 3000 rpm for 20 s. The glass was pre-baked at 110 °C for 90 s. The resist was exposed to the integral amount of 250 mJ/cm^2 (contact mode) and developed in NMD-3 for 10 min (Figure 2p). The glass was rinsed with DI water for 2 min and post-baked at 140 °C for 5 min. A Ti film (10-nm thick film) and Pt film (100-nm thick film) were deposited on three glass slides for 1 min as an adhesion layer by RF magnetron sputtering (L-250S-FH, Anelva Co., Ltd., Tokyo, Japan) (Figure 2q). Pt/Ti liftoff was performed in acetone with an ultrasonic cleaner (100 kHz) for 20 min (Figure 2r). The surface was then cleaned with IPA and nitrogen gas was blown off. The glass was drilled with a 1-mm diameter to form a hole for liquid introduction (Figure 2s). The ITO or Pt/Ti electrodes and PDMS micro-channels were bonded after air plasma (Figure 2t,u).

2.4. Observation Setup and Voltage Application

An inverted microscope (ECLIPSE Ti-U, Nikon Co., Ltd., Tokyo, Japan), equipped with ×4, ×10, and ×20 objective lenses, and a cooled charge-coupled device (CCD) camera (DS-Qi1Mc, Nikon) were used for flow observation. The movement of particles or cells transported with electrokinetic micropumps at 13–15 fps was recorded.

Voltage was applied to the interdigital electrodes by connecting it to a function generator (Protek 9305, GS Instruments Co., Ltd., Incheon, Korea) via an Ag/AgCl electrode. One electrode was set to ground (GND) and the other electrode was set to a signal electrode. We applied AC voltage with a DC bias voltage V_{DC} in the range of +4 V to −4 V at a frequency of 100 kHz and a peak value V_{pp} of 8 V to 12 V.

2.5. Flow Characterization of Electrokinetic Flow Pump

The DC-biased ACEK flow was characterized as having ITO or Pt/Ti electrodes and a low conductivity isotonic solution buffer (8.5 w/v% sucrose and 0.3 w/v% glucose) [43] was used to suppress the influence of Joule heating in the solution. Fluorescent particles (Sigma, L4530, particle diameter 2.0 μm, concentration 2.86 × 10^7 particles/mL) were suspended in the low conductivity buffer and were used as a tracer. The solution was introduced into the flow channel with a syringe. After the introduction, the hole was covered with a glass coverslip to reduce the influence of flow generated by a pressure imbalance.

The particles were transported in the flow channel of the electrokinetic pump to evaluate their driving characteristics. We traced 10 particles from 10 s after from the start of voltage application and measured their moving distance for 30 frames (about 2.2 s). The average particle velocity was derived from the movement of the particles and was considered as flow velocity. The change in the electrokinetic flow velocity was investigated by changing voltage conditions.

2.6. Manipulation of Cells Using a Micro-Nozzle Array Integrated with an Electrokinetic Flow Pump

We generated a DC-biased ACEK flow and manipulated cells through the fabricated micro-nozzle array. An overview of the experiment setup is shown in Figure 3. The micro-nozzle array was fixed to XYZ stages via a frame. We adjusted the position on the XYZ stage so that the nozzle surface was close to the bottom of the Petri dish. The nozzle array was placed in a cell suspension. The electrodes were connected to the function generator and GND. HeLa cells (RIKEN, RCB0007 provided by the RIKEN BioResource Research Center through the National BioResource Project of the MEXT/AMED, Tsukuba, Japan), were suspended in a low conductivity buffer containing φ2.0 μm tracer particles. The cell concentration was adjusted to 4.28 × 10^5 cells/mL and the cell suspension was poured in a petri dish. The nozzle array was brought close to the bottom of the petri dish. A DC-biased AC signal was applied to the electrodes connected to one nozzle to generate an electrokinetic flow, and cells through single nozzles were manipulated. The applied signal was V_{DC} −4 V, frequency of 100 kHz, and V_{pp} 10 V.

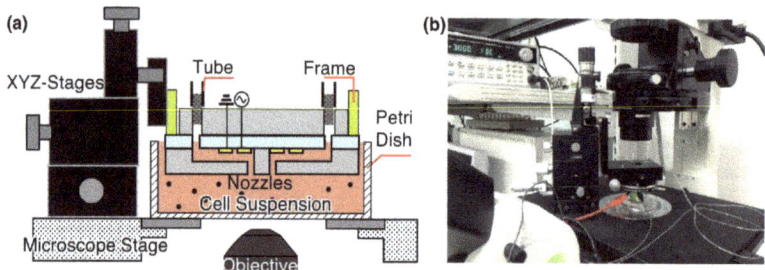

Figure 3. The experimental setup for cell manipulation with a developed nozzle array. (**a**) Schematic and (**b**) picture of the setup. A nozzle array was mounted on the XYZ stage via a 15 mm × 15 mm stainless frame. Electrodes were connected with AC power sources and ground (GND). In the channel of a nozzle array, particle suspension was introduced and HeLa cells were suspended in a petri dish.

3. Results and Discussion

3.1. Fabrication of the Integrated Nozzle Array and Pump

An array of 4 × 4 micronozzles integrated with electrokinetic flow pumps was fabricated (Figure 4, Table 1). Each part was fabricated by molding, lift-off, and etching. The nozzle array integrated with pumps were made of transparent PDMS and ITO electrodes, which enabled observation of the inside of a channel. Additionally, 4 × 4 nozzles were opened as water passed through the nozzles. The nozzle's inner diameter was 34.8 ± 1.04 μm (N = 16, mean ± standard deviation) (Figure 4a,b) and enlarged from 30 μm. PDMS micro-nozzles were connected to flow channels and ITO electrodes were patterned on a glass substrate (Figure 4c). The electrokinetic flow pump consisted of seven pairs of electrodes. The ITO interdigital electrodes were arranged in the flow channel with a width of 71.9 μm, a gap of 27.8 μm, and a pitch of 278.8 μm. The flow channel and pump were separated from each other by the PDMS wall, and each pump had its own signal input electrode. This design enabled individual flow control. The advantage of employing electrical pumps for cell transport was that it provided true scalability to one-side-open multi-channel devices.

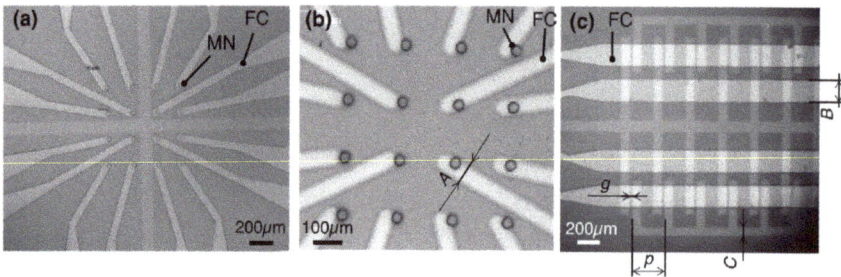

Figure 4. Micrographs of micro-nozzles integrated with ITO micropumps. (**a**) 4 × 4 micro-nozzles connected with micro-channels. Nozzle and channel were fabricated in the transparent materials PDMS and glass, which enabled observation of any transmission. (**b**) Close-up view of a micro-nozzle array. (**c**) Four series of flow channels and electrodes. FC and MN denote flow channel and a micro-nozzle, respectively. Typical dimensions were measured from five points: Nozzle diameter 34.8 μm. Channel width, A = 62.0 μm. B = 189.1 μm. Height, H = 71.9 μm. Electrode width, C = 71.9 μm. Gap g = 27.8 μm. Pitch p = 278.8 μm.

Table 1. Dimensions of the micro-nozzles integrated with indium tin oxide (ITO) micropumps and electrodes.

Category	Parts	Dimension
Flow channel	Height, H	71.9 ± 1.41 μm (N = 16)
	Width of flow channel in electrode section, B	189.1 ± 4.57 μm (N = 10)
	Width of the narrower flow channel, A	62.0 ± 2.30 μm (N = 16)
ITO electrodes	Width, C	65.7 ± 0.91 μm (N = 5)
	Gap between a pair of electrodes, g	23.3 ± 1.81 μm (N = 5)
	Pitch, p	191.0 ± 1.11 μm (N = 5).
Pt/Ti electrodes	Width, C	67.8 ± 0.91 μm (N = 5)
	Gap between a pair of electrodes, g	24.5 ± 1.44 μm (N = 5)
	Pitch, p	192.7 ± 1.12 μm (N = 5)

ITO or Pt/Ti micropumps with single-layer micro-channels and a 0.5-mm hole for characterization of DC-biased ACEKPs were also fabricated. Both the electrodes were arranged at regular intervals.

3.2. Characterization of DC-Biased ACEK Flow Using Electrokinetic Flow Pump

DC bias was switched to positive or negative for generation of a bidirectional flow. Figure 5 shows the transport of particles (φ2 μm) in a low conductive isotonic buffer at V_{pp} 10 V, 100 kHz, and V_{DC} ± 3 V. Unless stated, V_{pp} 10 V and 100 kHz were used. The application of positive and negative V_{DC} generated a net flow from the signal electrode to the ground electrode and vice versa. The voltage application of V_{DC} +3 V and −3 V moved particles from the electrode side to the nozzle side (Figure 5b, Movie S1) and from the nozzle side to the electrode side (Figure 5c, Movie S2). The application of voltage immediately created flow and the shut-down of voltage application stopped the generated flow. Movie S3 presents a vortex flow near a pair of ITO electrodes, and the creation of a net bidirectional flow through a PDMS micro-channel.

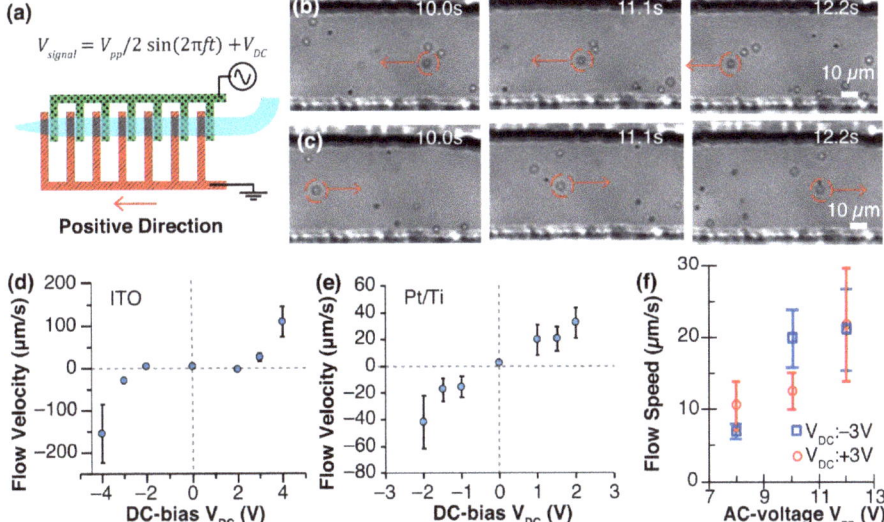

Figure 5. Generation of DC-biased AC electrokinetic flow with an ITO or Pt/Ti pump. The frequency and peak-to-peak voltage of the AC signal were set to 100 kHz and V_{pp} 10 V, respectively. (a) Schematic diagram of the pump. Particle transportation with an ITO pump at (b) V_{DC} +3 V and (c) V_{DC} −3 V. Electrokinetic flow velocity versus DC bias with (d) ITO and (e) Pt/Ti electrodes. (f) Flow speed versus V_{pp} voltage with ITO electrodes. Flow velocity was measured from 10 particles moving in the channel.

The average flow velocity was obtained according to the bias voltage (V_{DC}), where the direction from the signal electrode to the ground electrode was defined as positive (Figure 5d–f). The flow speed increased with an increase in the applied DC bias at both the ITO electrode and the Pt/Ti electrode (Figure 5d,e). With the ITO pump, the flow velocities were positive at V_{DC} +3 V or greater, and negative at V_{DC} −3 V or less. The flow velocities were 108.7 ± 36.4 µm/s (N = 10) at V_{DC} + 4 V and −155.6 ± 69.5 µm/s (N = 10) at V_{DC} − 4 V. We calculated the maximum flow rate of the pump with the product of a flow speed of 156 µm/s and a cross-section area of 4.46×10^3 µm^2 (channel width of 62.0 µm × channel height of 71.9 µm). The flow rate of the pump was 6.96×10^5 µm^3/s, which was equal to 696 pL/s. The flow velocities are almost symmetrical about the origin, but there is some difference. A two-electrode system was employed for flow generation because of its simplicity, while the potential of the electrodes was difficult to control. The addition of reference electrodes and a three-electrode system would stabilize the potential of the electrodes and flow generation. Bubbles were generated between Pt/Ti electrodes at V_{DC} ±3 V and bridged a pair of the electrodes after 30 s of voltage application. Therefore, the flow velocities at V_{DC} = ±3 V and ±4 V were not measured. The comparison of electrode materials shows that ITO pumps generate a faster flow than Pt/Ti electrodes and are suitable for large-volume fluid pumping. Pt/Ti electrodes are more stable than ITO and satisfy a long period of usage.

While most of the cell dimensions are between 10–20 µm, we prioritized the measurement of the fluid velocity and chose φ2-µm tracer particles in our characterization experiments. The smaller particles decreased the inertia of tracers and the measurement error [44]. The smaller size particles improve the response to flow, whereas the response of HeLa cells is slower than the particles. The suction and release of cells are assumed to take more time than the results characterized with φ2-µm particles. The particle concentration was 2.86×10 particles/nL and this concentration was low enough to not cause interactions between particles.

ITO pumps did not generate a significant electrokinetic flow in the range of V_{DC} of −2 V to +2 V. The Pt/Ti pumps generated a significant electrokinetic flow at V_{DC} of ±1 V. While a bubble formed between a pair of Pt/Ti electrodes at V_{DC} of ±3 V, bubbles did not form between a pair of ITO electrodes at V_{DC} of ±3 V. Bubble generation relates to the electrolysis of the solution. The reason why ITO pumps did not create a clear electrokinetic flow with a small DC bias is because the voltage did not go above the overpotential of water electrolysis. The working principle of the DC-biased AC pump was assumed to be based on a transverse conductivity gradient. This gradient was created through incipient Faradaic reactions occurring at the electrodes when a DC-bias AC voltage was applied to gold electrodes [38]. The electrolytic decomposition reaction of water occurred, as shown below. H$^+$ ions were generated at the anode side electrode.

$$2H_2O(L) \rightarrow 4H^+ (aq) + O_2(g) + 4e^- \quad (1)$$

OH$^-$ ions are generated from the cathode side electrode.

$$4H_2O(L) + 4e^- \rightarrow 4OH^- (aq) + 2H_2(g) \quad (2)$$

Because the difference in conductivity between H$^+$ and OH$^-$ was made in these reactions, a conductivity gradient was created in the solution on the anode and the cathode. This conductivity gradient and voltage application generated vortex flow between symmetrical electrodes. The change of electrode materials affected the overpotential of water electrolysis.

Peak-to-peak amplitude, V_{pp} on speed of DC-biased ACEK flow were investigated. With increasing V_{pp}, flow speeds increased (Figure 5f). Absolute flow speeds were plotted for ease of comparison. Flow speed was around 10 µm/s at V_{pp} 9 V and around 20 µm/s at V_{pp} 11 V for positive and negative DC bias. Based on the above observations, it was concluded that the larger the absolute value of the DC bias voltage or AC voltage, the greater the flow speed.

3.3. Cell Manipulation by Pump-Integrated Micro-Nozzle Array

We manipulated 10 cells in a parallel manner by electrokinetic flow with three of 16 nozzles (Figure 6, Movie S4). HeLa cells were suspended in the low conductive isotonic buffer and poured in the bottom of a Petri dish. A nozzle array was immersed in the cell suspension with mechanical stages. Before suction, cells distributed everywhere in the field. Since the largest flow velocity was obtained with ITO electrodes at V_{DC} ±4 V from the previous section, DC-biased AC voltages (V_{DC} −4 V, V_{pp} 10 V, 100 kHz) was applied and the pump generated an inward flow to the flow channel. During the application of the voltage, a suction flow was continuously created and vortex flow was generated in between a pair of electrodes. We focused on three nozzles for characterization rather than 16 nozzles due to the limitation of the observation area. When the gap between the nozzle array and the bottom of the dish was about 50 µm, only φ2-µm particles were sucked and cells were not sucked into the nozzles for 120 s. The gap was further reduced to around 20–30 µm and cells were transported. A cell denoted as #2 in Figure 6 was aspirated into the middle nozzle at about 76.6 s after the start of the application (Figure 6c). Each cell was attracted to the nearest nozzle (Figure 6d). These results indicate that the liquid volume should be reduced to create a sufficient flow for cell transportation. The nozzle array was fabricated in transparent materials such as PDMS, glass, and ITO, so that the structure did not prevent observation of the cells.

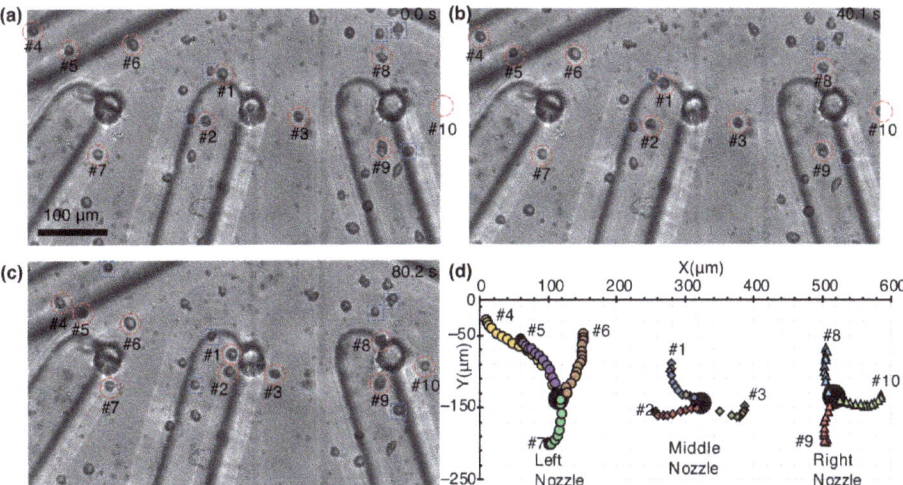

Figure 6. Suction of 10 cells using an electrokinetic pump near three nozzles (100 kHz, 10 V_{pp}, V_{DC} −4 V). (**a**–**c**) Time-lapse images of cell suction. Voltage was applied at 0 s and the flow was continuously generated. The 10 cells are circled by a red dashed line and attracted cells are boxed by a blue dashed line. (**d**) Trajectories of the aspirated 10 cells.

Ten cells near the nozzles were sucked into the nozzles during 3-min pumping. The throughput of cell suction was 1.1 ± 0.2 cells/min (N = 3) for each nozzle. Their distances and speed were measured over time (Figure 7) and the distance of cells from each nozzle were plotted in Figure 7a. Successful cell suction took place within a distance of 150 µm from each nozzle. The other nearby seven cells (two cells on the left, two cells on the middle, and three cells on the right) were attracted to each nozzle, but they did not reach the nozzles and their trajectories were not measured. The 10 cells travelled an average distance of 75.5 ± 26.2 µm (N = 10) and the average aspiration speed was 0.69 ± 0.19 µm/s (N = 10). A cell denoted as #4 in Figures 6 and 7 showed the longest travel distance of 151 µm. The speed of cells increased over time (Figure 7b). The cell speed was around 0.5 µm/s in the initial position and around 1 µm/s in the adjacent nozzles. For more rapid cell manipulation, the manipulation speed needs to be increased. One solution is to increase the number of electrode pairs from seven pairs.

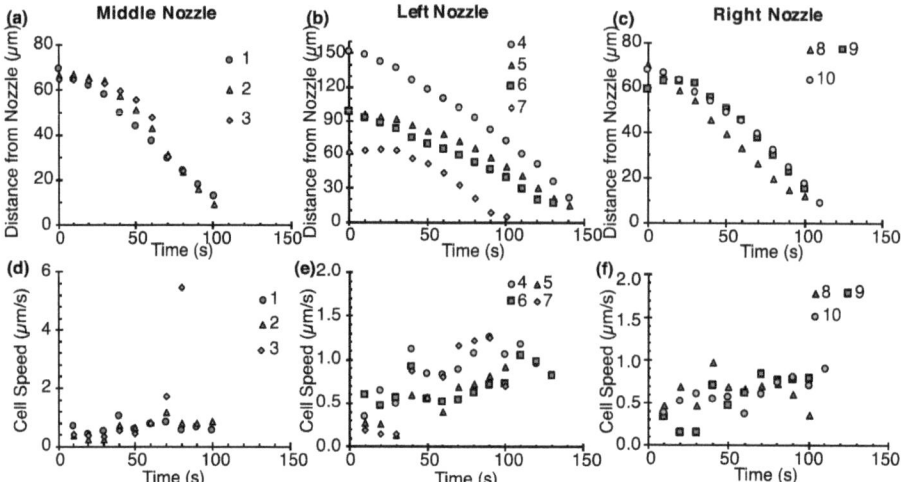

Figure 7. Characterization of parallel single cell suction. (**a**–**c**) Relationship between the distance of single cells from the nearest nozzle and time. (**d**–**f**) Relationship between the speed of the cells and time.

The cells were transported into the flow channels outside the nozzles. The shut-down of the voltage application stopped the movement of the cells. Three of the 10 cells were flowed through the channels at 3.39 ± 1.65 µm/s (N = 3). The transporting speed of the cells in the channel was much slower than the average flow speed obtained from the previous section. This is because the flow resistance of a micro-nozzle is two orders of magnitudes higher than that of a punched hole. Fluidic resistance through a circular pipe, R, is expressed as:

$$R = \frac{128}{\pi} \eta L \frac{1}{D^4}$$

where η is the dynamic viscosity of the liquid, L is the channel length, and D is the circle diameter [45]. The ratio of hydraulic resistances of a nozzle and hole can be written as $R_{nozzle}/R_{hole} = L_{nozzle}/L_{hole}(D_{hole}/D_{nozzle})^4$. Inserting D_{nozzle} = 35 µm, L_{nozzle} = 70 µm, D_{hole} = 0.5 mm, and L_{hole} = 8 mm into the equation yields R_{nozzle}/R_{hole} = 364.

There is potential risk for cell damage associated with electrical field and mechanical stress. When electric field strength is above the electroporation threshold (0.4–0.6 kV/cm), electroporation occurs in the cell and the cell membrane becomes permeable [46–48]. In this study, however, the voltage was applied only to the electrodes in the channel, which was far from nozzles. Cells were manipulated without reaching electrodes and their motions were stopped by turning off the voltage. Cells did not need to pass through a high electric field near a pair of electrodes. Therefore, electrokinetic forces on cells can be ignored. Suspended cells were not largely deformed during manipulation and experienced mainly Stokes' drag in a low Reynolds regime [49]. Since cell speeds were low, we thought strong mechanical stress due to Stokes' drag was not applied to cells during cell manipulation.

Although it is possible to control the flow through a single nozzle electrically, the bottleneck in large-scale single cell manipulation is the number of signal generators. Analog demultiplexer [50] enables a decrease in the number of signal generating circuits and to maintain the controllability of multiple channels. The placement of trapped single cells in microwells can be achieved with mechanical stages and a microscope [21,30].

4. Conclusions

In this study, 4 × 4 micronozzles were integrated with microchannels and electrodes for parallel manipulation of single cells. A DC-biased AC signal was applied to the interdigital tooth electrode made of ITO or Pt/Ti, and an electrokinetic flow occurred at both electrodes in a low conductivity isotonic buffer. By switching the DC bias voltages from positive to negative, the direction of flow was controlled. Since the electrokinetic flow was not significant at small DC bias on ITO electrodes, there was a threshold value of the voltage for generating the flow. Using a micro-nozzle enabled the generation of flow inward and outward of the micro-nozzle and cell manipulation. A flow generated with the electrokinetic pump transported cells into the micro-nozzle array and it was possible to demonstrate manipulation of single cells using the DC-biased ACEK flow.

The developed method of integrating nozzles and manipulation method is fundamental for creating high-throughput single manipulation tools accessible to an open-top chip. Our integrated system collected single cells at 1 event/min, whereas an on-demand droplet collection system dispensed single cells at a throughput of 20 events/min [29]. The improvement of the throughput and flow rate required an increase in the number of interdigital-tooth electrodes. The true value of this study is that this integration technology provides individual flow control of each micro-channel and scalability with one-side-open multi-channel devices [29,30]. Our future works includes a viability assay and rearrangement of single cells with the nozzles integrated with pumps.

Supplementary Materials: The following are available online at http://www.mdpi.com/2072-666X/11/4/442/s1. Movie S1, S2: Movement of particles ($\varphi 2$ μm) in a low conductive isotonic buffer at V_{pp} 10 V, 100 kHz. Movie S1: V_{DC} + 3 V and Movie S2: −3 V. Voltage is not applied for 5 s from the start of the movie. Movie S3: Vortex flow near a pair of ITO electrodes, and creation of net flow through a PDMS microchannel. Movie S4: Suction of 10 cells to three micro-nozzles integrated with ITO pumps at V_{DC} −4 V, V_{pp} 10 V, and 100 kHz.

Author Contributions: Conceptualization, M.N. Methodology, M.N., K.K., and S.S. Validation, M.N., K.K., and S.S. Writing—original draft preparation, M.N. and K.K. Writing—review and editing, T.S.S. and T.S. Supervision, M.N. and T.S. Project administration, M.N. Funding acquisition, M.N. All authors have read and agreed to the published version of the manuscript.

Funding: The JSPS KAKENHI (Grant number: 25820087, 15K13910, and 16H06074), the Tokai Foundation for Technology, and Adaptable and Seamless Technology Transfer Program through Target-driven R&D (A-STEP, VP30218089109) from Japan Science and Technology Agency (JST) supported this work.

Acknowledgments: We acknowledge Kiyotaka Oohara for technical assistance with the experiments.

Conflicts of Interest: The authors declare no conflict of interest.

References

1. Bhatia, S.; Balis, U.; Yarmush, M.; Toner, M. Microfabrication of hepatocyte/fibroblast co-cultures: Role of homotypic cell interactions. *Biotechnol. Prog.* **1998**, *14*, 378–387. [CrossRef] [PubMed]
2. Hofmann, A.; Ritz, U.; Verrier, S.; Eglin, D.; Alini, M.; Fuchs, S.; Kirkpatrick, C.J.; Rommens, P.M. The effect of human osteoblasts on proliferation and neo-vessel formation of human umbilical vein endothelial cells in a long-term 3D co-culture on polyurethane scaffolds. *Biomaterials* **2008**, *29*, 4217–4226. [CrossRef] [PubMed]
3. Shen, Q.; Goderie, S.K.; Jin, L.; Karanth, N.; Sun, Y.; Abramova, N.; Vincent, P.; Pumiglia, K.; Temple, S. Endothelial cells stimulate self-renewal and expand neurogenesis of neural stem cells. *Science* **2004**, *304*, 1338–1340. [CrossRef] [PubMed]
4. Kaji, H.; Camci-Unal, G.; Langer, R.; Khademhosseini, A. Engineering systems for the generation of patterned co-cultures for controlling cell–cell interactions. *Biochim. Biophys. Acta (BBA) Gen. Subj.* **2011**, *1810*, 239–250. [CrossRef]
5. Hassanzadeh-Barforoushi, A.; Shemesh, J.; Farbehi, N.; Asadnia, M.; Yeoh, G.H.; Harvey, R.P.; Nordon, R.E.; Warkiani, M.E. A rapid co-culture stamping device for studying intercellular communication. *Sci. Rep.* **2016**, *6*, 35618. [CrossRef]
6. Shinde, P.; Mohan, L.; Kumar, A.; Dey, K.; Maddi, A.; Patananan, A.N.; Tseng, F.G.; Chang, H.Y.; Nagai, M.; Santra, T.S. Current Trends of Microfluidic Single-Cell Technologies. *Int. J. Mol. Sci.* **2018**, *19*, 3143. [CrossRef]

7. Skelley, A.M.; Kirak, O.; Suh, H.; Jaenisch, R.; Voldman, J. Microfluidic control of cell pairing and fusion. *Nat. Methods* **2009**, *6*, 147–152. [CrossRef]
8. Frimat, J.-P.; Becker, M.; Chiang, Y.-Y.; Marggraf, U.; Janasek, D.; Hengstler, J.G.; Franzke, J.; West, J. A microfluidic array with cellular valving for single cell co-culture. *Lab Chip* **2011**, *11*, 231–237. [CrossRef]
9. Hong, S.; Pan, Q.; Lee, L.P. Single-cell level co-culture platform for intercellular communication. *Integr. Biol.* **2012**, *4*, 374–380. [CrossRef]
10. Chen, Y.-C.; Cheng, Y.-H.; Kim, H.S.; Ingram, P.N.; Nor, J.E.; Yoon, E. Paired single cell co-culture microenvironments isolated by two-phase flow with continuous nutrient renewal. *Lab Chip* **2014**, *14*, 2941–2947. [CrossRef]
11. Bhardwaj, R.; Gupta, H.; Pandey, G.; Ryu, S.; Shibata, T.; Santra, T.S.; Nagai, M. Single-Cell Manipulation. In *Handbook of Single Cell Technologies*; Santra, T.S., Tseng, F.-G., Eds.; Springer Singapore: Singapore, 2020; pp. 1–26.
12. Gray, D.S.; Tan, J.L.; Voldman, J.; Chen, C.S. Dielectrophoretic registration of living cells to a microelectrode array. *Biosens. Bioelectron.* **2004**, *19*, 771–780. [CrossRef] [PubMed]
13. Albrecht, D.R.; Tsang, V.L.; Sah, R.L.; Bhatia, S.N. Photo-and electropatterning of hydrogel-encapsulated living cell arrays. *Lab Chip* **2005**, *5*, 111–118. [CrossRef] [PubMed]
14. Park, H.; Kim, D.; Yun, K.-S. Single-cell manipulation on microfluidic chip by dielectrophoretic actuation and impedance detection. *Sens. Actuators B: Chem.* **2010**, *150*, 167–173. [CrossRef]
15. Chiou, P.Y.; Ohta, A.T.; Wu, M.C. Massively parallel manipulation of single cells and microparticles using optical images. *Nature* **2005**, *436*, 370–372. [CrossRef]
16. Juan, M.L.; Righini, M.; Quidant, R. Plasmon nano-optical tweezers. *Nat. Photonics* **2011**, *5*, 349–356. [CrossRef]
17. Mirsaidov, U.; Scrimgeour, J.; Timp, W.; Beck, K.; Mir, M.; Matsudaira, P.; Timp, G. Live cell lithography: Using optical tweezers to create synthetic tissue. *Lab Chip* **2008**, *8*, 2174–2181. [CrossRef]
18. Tumarkin, E.; Tzadu, L.; Csaszar, E.; Seo, M.; Zhang, H.; Lee, A.; Peerani, R.; Purpura, K.; Zandstra, P.W.; Kumacheva, E. High-throughput combinatorial cell co-culture using microfluidics. *Integr. Biol.* **2011**, *3*, 653–662. [CrossRef]
19. Hassanzadeh-Barforoushi, A.; Law, A.M.K.; Hejri, A.; Asadnia, M.; Ormandy, C.J.; Gallego-Ortega, D.; Ebrahimi Warkiani, M. Static droplet array for culturing single live adherent cells in an isolated chemical microenvironment. *Lab Chip* **2018**, *18*, 2156–2166. [CrossRef]
20. Wilson, C.F.; Simpson, G.J.; Chiu, D.T.; Strömberg, A.; Orwar, O.; Rodriguez, N.; Zare, R.N. Nanoengineered structures for holding and manipulating liposomes and cells. *Anal. Chem.* **2001**, *73*, 787–791. [CrossRef]
21. Nagai, M.; Kato, K.; Oohara, K.; Shibata, T. Pick-and-Place Operation of Single Cell Using Optical and Electrical Measurements for Robust Manipulation. *Micromachines* **2017**, *8*, 350. [CrossRef]
22. Han, H.; Martinez, V.; Aebersold, M.J.; Lüchtefeld, I.; Polesel-Maris, J.; Vörös, J.; Zambelli, T. Force controlled SU-8 micropipettes fabricated with a sideways process. *J. Micromech. Microeng.* **2018**, *28*, 095015. [CrossRef]
23. Martinez, V.; Forró, C.; Weydert, S.; Aebersold, M.J.; Dermutz, H.; Guillaume-Gentil, O.; Zambelli, T.; Vörös, J.; Demkó, L. Controlled single-cell deposition and patterning by highly flexible hollow cantilevers. *Lab Chip* **2016**, *16*, 1663–1674. [CrossRef] [PubMed]
24. Hunt, T.; Westervelt, R. Dielectrophoresis tweezers for single cell manipulation. *Biomed. Microdevices* **2006**, *8*, 227–230. [CrossRef] [PubMed]
25. Kodama, T.; Osaki, T.; Kawano, R.; Kamiya, K.; Miki, N.; Takeuchi, S. Round-tip dielectrophoresis-based tweezers for single micro-object manipulation. *Biosens. Bioelectron.* **2013**, *47*, 206–212. [CrossRef] [PubMed]
26. Yusof, A.; Keegan, H.; Spillane, C.D.; Sheils, O.M.; Martin, C.M.; O'Leary, J.J.; Zengerle, R.; Koltay, P. Inkjet-like printing of single-cells. *Lab Chip* **2011**, *11*, 2447–2454. [CrossRef] [PubMed]
27. Gross, A.; Schondube, J.; Niekrawitz, S.; Streule, W.; Riegger, L.; Zengerle, R.; Koltay, P. Single-cell printer: Automated, on demand, and label free. *J. Lab. Autom.* **2013**, *18*, 504–518. [CrossRef]
28. Cole, R.H.; Tang, S.-Y.; Siltanen, C.A.; Shahi, P.; Zhang, J.Q.; Poust, S.; Gartner, Z.J.; Abate, A.R. Printed droplet microfluidics for on demand dispensing of picoliter droplets and cells. *Proc. Natl. Acad. Sci. USA* **2017**, *114*, 8728–8733. [CrossRef]
29. Nan, L.; Lai, M.Y.A.; Tang, M.Y.H.; Chan, Y.K.; Poon, L.L.M.; Shum, H.C. On-Demand Droplet Collection for Capturing Single Cells. *Small* **2020**, *16*, 1902889. [CrossRef]

30. Nagai, M.; Oohara, K.; Kato, K.; Kawashima, T.; Shibata, T. Development and characterization of hollow microprobe array as a potential tool for versatile and massively parallel manipulation of single cells. *Biomed. Microdevices* **2015**, *17*, 41. [CrossRef]
31. Iverson, B.D.; Garimella, S.V. Recent advances in microscale pumping technologies: A review and evaluation. *Microfluid. Nanofluid.* **2008**, *5*, 145–174. [CrossRef]
32. Wang, X.Y.; Cheng, C.; Wang, S.L.; Liu, S.R. Electroosmotic pumps and their applications in microfluidic systems. *Microfluid. Nanofluid.* **2009**, *6*, 145–162. [CrossRef] [PubMed]
33. Shibata, T.; Nakamura, K.; Horiike, S.; Nagai, M.; Kawashima, T.; Mineta, T.; Makino, E. Fabrication and characterization of bioprobe integrated with a hollow nanoneedle for novel AFM applications in cellular function analysis. *Microelectron. Eng.* **2013**, *111*, 325–331. [CrossRef]
34. Nagai, M.; Torimoto, T.; Miyamoto, T.; Kawashima, T.; Shibata, T. Electrokinetic Delivery of Biomolecules into Living Cells for Analysis of Cellular Regulation. *Jpn. J. Appl. Phys.* **2013**, *52*, 047002. [CrossRef]
35. Cahill, B.P.; Heyderman, L.J.; Gobrecht, J.; Stemmer, A. Electro-osmotic streaming on application of traveling-wave electric fields. *Phys. Rev. E* **2004**, *70*, 036305. [CrossRef]
36. Ramos, A.; Morgan, H.; Green, N.G.; Gonzalez, A.; Castellanos, A. Pumping of liquids with traveling-wave electroosmosis. *J. Appl. Phys.* **2005**, *97*, 084906. [CrossRef]
37. Islam, N.; Reyna, J. Bi-directional flow induced by an AC electroosmotic micropump with DC voltage bias. *Electrophoresis* **2012**, *33*, 1191–1197. [CrossRef]
38. Ng, W.Y.; Ramos, A.; Lam, Y.C.; Wijaya, I.P.M.; Rodriguez, I. DC-biased AC-electrokinetics: A conductivity gradient driven fluid flow. *Lab Chip* **2011**, *11*, 4241–4247. [CrossRef]
39. Ng, W.Y.; Ramos, A.; Lam, Y.C.; Rodriguez, I. Numerical study of dc-biased ac-electrokinetic flow over symmetrical electrodes. *Biomicrofluidics* **2012**, *6*, 012817. [CrossRef]
40. Lian, M.; Wu, J. Ultrafast micropumping by biased alternating current electrokinetics. *Appl. Phys. Lett.* **2009**, *94*, 064101. [CrossRef]
41. Zhang, M.; Wu, J.; Wang, L.; Xiao, K.; Wen, W. A simple method for fabricating multi-layer PDMS structures for 3D microfluidic chips. *Lab Chip* **2010**, *10*, 1199–1203. [CrossRef]
42. Nagai, M.; Oguri, M.; Shibata, T. Characterization of light-controlled Volvox as movable microvalve element assembled in multilayer microfluidic device. *Jpn. J. Appl. Phys.* **2015**, *54*, 067001. [CrossRef]
43. Wu, L.; Lanry Yung, L.Y.; Lim, K.M. Dielectrophoretic capture voltage spectrum for measurement of dielectric properties and separation of cancer cells. *Biomicrofluidics* **2012**, *6*, 014113. [CrossRef] [PubMed]
44. Mei, R. Velocity fidelity of flow tracer particles. *Exp. Fluids* **1996**, *22*, 1–13. [CrossRef]
45. Bruus, H. *Theoretical Microfluidics*; Oxford University Press: Oxford, UK, 2008; p. 346.
46. Kranjc, M.; Miklavčič, D. Electric Field Distribution and Electroporation Threshold. In *Handbook of Electroporation*; Miklavčič, D., Ed.; Springer International Publishing: Cham, Switzerland, 2017; pp. 1043–1058.
47. Santra, T.S.; Chang, H.Y.; Wang, P.C.; Tseng, F.G. Impact of pulse duration on localized single-cell nano-electroporation. *Analyst* **2014**, *139*, 6249–6258. [CrossRef] [PubMed]
48. Santra, T.S.; Wang, P.-C.; Chang, H.-Y.; Tseng, F.-G. Tuning nano electric field to affect restrictive membrane area on localized single cell nano-electroporation. *Appl. Phys. Lett.* **2013**, *103*, 233701. [CrossRef]
49. Huber, D.; Oskooei, A.; Solvas, X.C.I.; DeMello, A.; Kaigala, G.V. Hydrodynamics in Cell Studies. *Chem. Rev.* **2018**, *118*, 2042–2079. [CrossRef]
50. Scherz, P.; Monk, S. *Practical Electronics for Inventors*, 4th ed.; McGraw-Hill Education: New York, NY, USA, 2016.

© 2020 by the authors. Licensee MDPI, Basel, Switzerland. This article is an open access article distributed under the terms and conditions of the Creative Commons Attribution (CC BY) license (http://creativecommons.org/licenses/by/4.0/).

Article

Characterization of Single-Nucleus Electrical Properties by Microfluidic Constriction Channel

Hongyan Liang [1,2,†], Yi Zhang [1,2,†], Deyong Chen [1,2], Huiwen Tan [1,2], Yu Zheng [3], Junbo Wang [1,2,*] and Jian Chen [1,2,*]

1. State Key Laboratory of Transducer Technology, Aerospace Information Research Institute, Chinese Academy of Sciences, Beijing 100094, China; lianghongyan17@mails.ucas.edu.cn (H.L.); zhangyi161@mails.ucas.edu.cn (Y.Z.); dychen@mail.ie.ac.cn (D.C.); tanhuiwen18@mails.ucas.edu.cn (H.T.)
2. School of Electronic, Electrical and Communication Engineering, University of Chinese Academy of Sciences, Beijing 101408, China
3. Shandong University, Jinan 250100, China; 201800301017@mail.sdu.edu.cn
* Correspondence: jbwang@mail.ie.ac.cn (J.W.); chenjian@mail.ie.ac.cn (J.C.); Tel.: +86-10-58887191 (J.W.); +86-10-58887256 (J.C.)
† Co-first authors.

Received: 27 September 2019; Accepted: 29 October 2019; Published: 31 October 2019

Abstract: As key bioelectrical markers, equivalent capacitance (C_{ne}, i.e., capacitance per unit area) and resistance (R_{ne}, i.e., resistivity multiply thickness) of nuclear envelopes have emerged as promising electrical indicators, which cannot be effectively measured by conventional approaches. In this study, single nuclei were isolated from whole cells and trapped at the entrances of microfluidic constriction channels, and then corresponding impedance profiles were sampled and translated into single-nucleus C_{ne} and R_{ne} based on a home-developed equivalent electrical model. C_{ne} and R_{ne} of A549 nuclei were first quantified as 3.43 ± 1.81 µF/cm² and 2.03 ± 1.40 Ω·cm² (N_n = 35), which were shown not to be affected by variations of key parameters in nuclear isolation and measurement. The developed approach in this study was also used to measure a second type of nuclei, producing C_{ne} and R_{ne} of 3.75 ± 3.17 µF/cm² and 1.01 ± 0.70 Ω·cm² for SW620 (N_n = 17). This study may provide a new perspective in single-cell electrical characterization, enabling cell type classification and cell status evaluation based on bioelectrical markers of nuclei.

Keywords: microfluidics; single-nucleus analysis; constriction channel; electrical properties; nuclear envelope

1. Introduction

Nuclear envelope defines a bilayer membrane that encloses the genome from the rest of the cell and regulates the movements of molecules across the nuclear-cytoplasmic boundary. As key bioelectrical markers, equivalent capacitance (C_{ne}, i.e., capacitance per unit area) and resistance (R_{ne}, i.e., resistivity multiply thickness) of the nuclear envelope have emerged as promising electrical indicators related to salivary gland cells with changes in developments [1–6], oocytes isolated from different species [7–13], normal and malignant white blood cells [14–20], yeasts with changes in extracellular solutions [21,22], epithelia cell lines from multiple sources [23–36] and differentiation of stem cells [37]. A summary of previously reported electrical parameters of single nuclei could be found in Table S1.

Patch clamping was initially adopted to characterize electrical properties of nuclear envelopes, where two glass pipettes were deployed within the nuclear and cytoplasmic domains, respectively, enabling the quantitation of C_{ne} and R_{ne} [1–9,23,38–40]. Based on this approach, C_{ne} and R_{ne} of single nuclei were characterized as 412 ± 62 µF/cm² and 1.5 ± 0.4 Ω·cm² [1], ~100 µF/cm² and 3.9 ± 1.4 Ω·cm² [2], 0.72 ± 0.09 Ω·cm² [4] and 2 Ω·cm² [6] of salivary gland cells of *Drosophila flavorepleta*.

Although powerful, patch clamping was mostly used to measure ultra large cells with diameters of nearly 100 μm and it is full of challenges to accurately penetrate nuclear domains of conventional eukaryotic cells with diameters around 10 μm.

In a second approach, two electrodes were inserted into cell suspensions and the corresponding impedance data were interpreted into both electrical parameters of cell membranes and nucleus [15, 17,18,21,24,25,28,29,31,34,36,41–45]. Based on this approach, C_{ne} and R_{ne} of nuclei derived from mouse spleen lymphocytes, human Jurkat cells and rat H9C2 cells were quantified as 0.62 μF/cm^2 and 0.07 Ω·cm^2 [15], 1.19 ± 0.14 μF/cm^2 and 0.21 ± 0.02 Ω·cm^2 [36], 0.22 ± 0.05 μF/cm^2 and 7.25 ± 0.68 Ω·cm^2 [34], respectively. However, this is an approach based on cell populations rather than single cells where potential concerns of interactions among neighboring cells during measurements cannot be properly addressed.

In order to address this issue, electrorotation was adopted for the characterization of single-nucleus electrical properties where single cells were forced to rotate by four electrodes and the rotating speeds as a function of the applied electrical signals were used to estimate electrical properties of nuclear envelopes [14,16,19,25,46]. However, in this approach, a double shell electrical model representing cell and nuclear membranes was used. Thus, capacitive properties of cell membranes may mask electrical properties of nuclear membranes at the low-frequency domain, while at the high-frequency domain, both cell membranes and nuclear membranes are short circuited.

To address this issue, in this study, cell lysis and nucleus isolation were first conducted and the obtained single nuclei were trapped at the entrance of the constriction channel for electrical property characterization (see Figure 1). In operation, trypsin-EDTA, the lysis buffer and IGEPAL@ CA-630 were used to acquire nuclei of cells (see Figure 1a). Then individual nucleus was trapped at the entrance of the constriction channel with corresponding impedance values measured by a lock-in amplifier (see Figure 1b). Based on an equivalent electrical model where nuclear electrical components were represented as C_{ne}, R_{ne} and R_{np} (equivalent resistance of nucleoplasm), raw impedance data were translated into C_{ne} and R_{ne} (see Figure 1c).

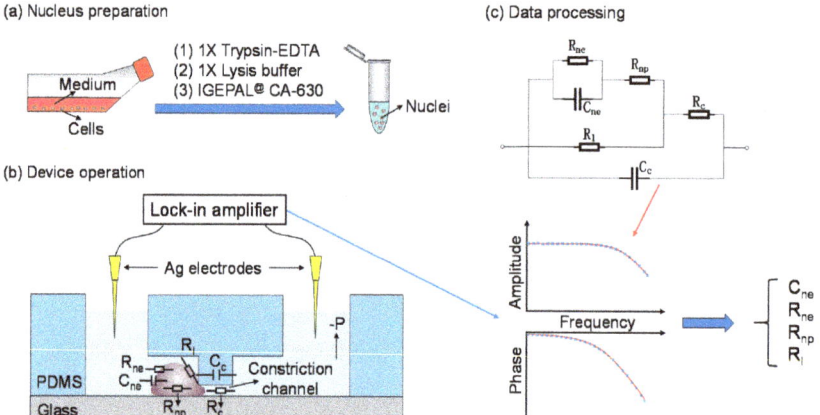

Figure 1. Working flowchart for characterizing single-nucleus electrical properties (e.g., C_{ne} and R_{ne}, two key parameters of nuclear envelope) based on microfluidic constriction channel. Key steps include nucleus preparation (**a**), device operation (**b**) and data processing (**c**). In operation, trypsin-EDTA (Ethylenediaminetetraacetic acid), lysis buffer and IGEPAL@ CA-630 were used to acquire nuclei of cells. Then single nuclei were trapped at the entrance of the constriction channel with corresponding impedance values measured by a lock-in amplifier. Based on an equivalent electrical model where nuclear electrical components were represented as C_{ne} in parallel with R_{ne}, and then in series with R_{np}, raw impedance data were translated into C_{ne} and R_{ne}.

In comparison to the aforementioned population approach for nuclear electrical property characterization, this constriction channel based approach enables the electrical property characterization at single nucleus level. In comparison to electrorotation, in this study, side effects of membrane capacitance on the estimation of nuclear electrical properties are addressed. In comparison to patch clamping, this device can easily trap single nuclei at the entrance of the constriction channel without the requirements of accurate manipulations of pipette tips.

2. Materials and Methods

2.1. Materials

All cell lines were purchased from China Infrastructure of Cell Line Resources. All reagents for cell culture (e.g., culture medium, fetal bovine serum and trypsin) were purchased from Life Technologies Corporation (Van Allen Way Carlsbad, CA, USA). All reagents for cell treatments (e.g., isotonic lysis buffer, dithiothreitol, protease inhibitor cocktail and IGEPAL® CA-630) were purchased from Sigma Aldrich Corporation (St. Louis, MO, USA). The materials required for device fabrication included SU-8 photoresist (MicroChem Corporation, Newton, MA, USA) and 184 silicone elastomer (Dow Corning Corporation, Midland, MI, USA).

2.2. Nucleus Preparation

A549 and SW620 cancer cell lines were cultured in RPMI 1640 supplemented with 10% fetal bovine serum, 1% penicillin and streptomycin, under the conditions of 37 °C and 5% CO_2. For nucleus preparation, it was adopted from a previous study [47] where the cultured cancer cells were trypsinized using 1x trypsin-EDTA from culture flasks, which were then incubated with an isotonic lysis buffer supplemented with 1 mM dithiothreitol and protease inhibitor cocktail on ice for 15 min. After that IGEPAL® CA-630 was added into the cell suspension, followed by a vortex for 10 s, and then a centrifugation of 400× g for 5 min. In the end, isolated nuclei were resuspended in 1x phosphate buffer saline containing 1% bovine serum albumin (see Figure 1a). Note that in nuclear isolation, variations of IGEPAL® CA-630 were used for comparison where 0.01%, 0.02% and 0.03% of IGEPAL® CA-630 were used for A549 cells while 0.01% of IGEPAL® CA-630 was used for SW620 cells.

2.3. Device Fabrication

The device mainly consists of constriction channels (cross-sectional dimensions of 7 µm × 8 µm for A549 nuclei or 5 µm × 5 µm for SW620 nuclei) in polydimethylsiloxane (PDMS) elastomers, which were replicated from SU-8 mold masters using conventional soft lithography. Briefly, SU-8 5 was spin-coated, exposed without development to form the layer of the constriction channel with a height of 7 µm or 5 µm. Then, SU-8 25 was spin coated, exposed with alignment and developed to form the nucleus-loading channel with a height of 25 µm. After fabricating SU-8 mold masters, PDMS precursor and curing agents (ratio 10:1 by weight) were thoroughly mixed, degassed and poured onto the SU-8 channel masters for crosslinking (4 hours at 80 °C). Fully cured PDMS channels were then peeled away, punched with through holes as inlets and outlets, and bonded to glass slides after plasma treatment.

2.4. Device Operation

In experiments, the microfabricated channels were first filled with 1× phosphate buffer saline containing 1% bovine serum albumin and then loaded with nuclei at a concentration of 1×10^6 nuclei/mL. A pressure controller (Pace 5000, Druck, Billerica, MA, USA) was used to generate negative pressures to trap single nuclei at the entrance of the constriction channels (0.2, 0.5 and 1.0 kPa for A549 nuclei at 7 µm × 8 µm constriction channels; 0.5, 1.0 and 2.0 kPa for SW620 nuclei at 5 µm × 5 µm constriction channels). Then, impedance profiles from 1 kHz to 250 kHz (excitation voltage: 200 mV) were recorded by a lock-in amplifier (7270, Signal Recovery, Oak Ridge, TN, USA). Meanwhile, an inverted microscope (IX83, Olympus, Tokyo, Japan) was used to capture the images of trapped single nuclei.

After characterization, high negative pressures were used to aspirate the nuclei through the constriction channels and then the device was ready for the next measurements (see Figure 1b).

2.5. Data Processing

To interpret the measured impedance data, an electrical model was proposed (see Figure 1c). The electrical model of the constriction channel was represented by an equivalent resistor (R_c) and a capacitor (C_c) in parallel. Nuclear electrical components were represented as a capacitor (C_{ne}) in parallel with a resistor (R_{ne}) for the portion of the nuclear envelope in series with a resistor (R_{np}) for the nucleoplasm portion. Furthermore, an equivalent leakage resistor (R_l) was defined to estimate sealing status between the aspirated membrane portion of the nucleus under the measurement and inner walls of the constriction channel.

Impedance profiles without nucleus trapping were first fitted with the equivalent electrical components of the constriction channel to obtain values of equivalent resistor (R_c) and capacitor (C_c). Then impedance profiles with nucleus trapping were fitted with the aforementioned electrical model, based on the nonlinear least-square principle, where a loop function was used to enumerate key parameters of C_{ne}, R_{ne}, R_{np} and R_l. Note that in the step of curve fitting of impedance profiles with nucleus trapping, R_c and C_c were treated as known variables without looping and thus the potential concern of parasitic capacitors of constriction channels on the extraction of electrical parameters of nuclear envelopes can be properly addressed.

2.6. Statistics

The measurements of multiple samples were conducted with results expressed by averages and standard deviations. The student's t-test was used, where the values of $p < 0.001$ (*) were considered as statistically significant.

3. Results and Discussion

Microfluidics refer to the manipulation of microscale fluids in microfabricated channels [48]. Due to dimensional comparison with cells, microfluidics has functioned as an enabling tool for single-cell isolation [49–52], and then impedance measurements [53]. Recently, single cells were trapped by a microfluidic device with corresponding impedance values measured, enabling the quantification of single-nucleus electrical properties [37]. However, in this approach, the electrical parameters of cell membranes can have side effects on the characterization of nuclear electrical properties. In this study, nuclei were isolated from cells and then trapped at the entrance of the constriction channel for electrical property characterization and thus the potential concern on the cell membranes can be properly addressed.

Figure 2 shows the microscopic images of isolated single nuclei stained with trypan blue, where stained blue dots and non-stained counterparts in the images represent isolated nuclei and intact cells, respectively. Under the conditions of 0.01%, 0.02% and 0.03% IGEPAL@ CA-630 for the treatments of A549 cells, it was observed that the increase of the percentage of IGEPAL@ CA-630 enhanced the ratio of cell lysis and percentage of nuclear isolation (see Figure 2a–c). As to SW620 cells, 0.01% IGEPAL@ CA-630 can produce comparable results with A549 cells under the treatment of 0.02% IGEPAL@ CA-630 (see Figure 2d).

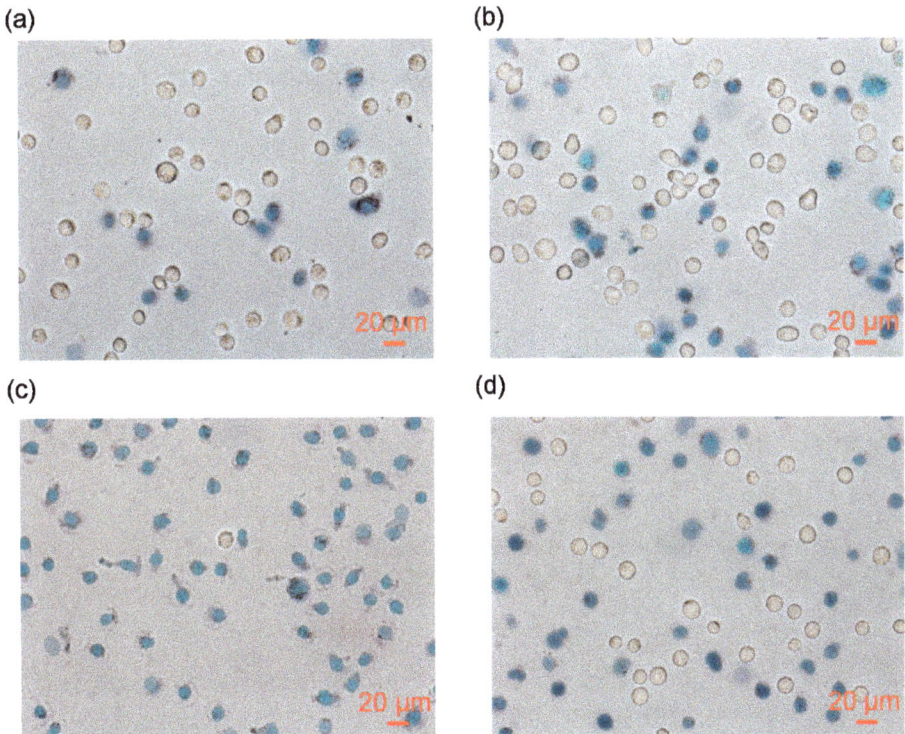

Figure 2. Microscopic images of isolated single nuclei stained with trypan blue, under the conditions of 0.01% IGEPAL® CA-630 for A549 cells (**a**), 0.02% IGEPAL® CA-630 for A549 cells (**b**), 0.03% IGEPAL® CA-630 for A549 cells (**c**) and 0.01% IGEPAL® CA-630 for SW620 cells (**d**). Note that stained blue dots and non-stained counterparts in the images represent isolated nuclei and intact cells, respectively.

Figure 3 shows measured impedance values with curve fitting and correspond microscopic images for trapped single nuclei at the entrance of the constriction channels. As shown in Figure 3a, for A549 nuclei, under the conditions of 0.01% IGEPAL® CA-630 and a constriction-channel of 7 μm × 8 μm, with the increase of aspiration pressure (P_a), impedance amplitude and phase values were noticed to increase and decrease, respectively. Meanwhile, the length of the nucleus aspirated into constriction channel extended.

Curve fitting (fit-base line) of measured impedance data without trapped single nuclei at the entrance of the constriction channel was conducted where C_c and R_c were quantified as 0.20 pF and 1.19 MΩ, respectively. Curve fitting (fit-0.2 kPa line, fit-0.5 kPa line and fit-1.0 kPa line) of measured impedance data with trapped single nuclei at the entrance of the constriction channel was conducted where C_{ne}, R_{ne}, R_{np} and R_l were quantified as 1.34 μF/cm², 4.48 Ω·cm², 0.50 MΩ and 0.45 MΩ (P_a = 0.2 kPa), 1.88 μF/cm², 1.85 Ω·cm², 0.60 MΩ and 0.80 MΩ (P_a = 0.5 kPa), 1.88 μF/cm², 3.72 Ω·cm², 0.60 MΩ and 0.95 MΩ (P_a = 1.0 kPa; see Figure 3a)

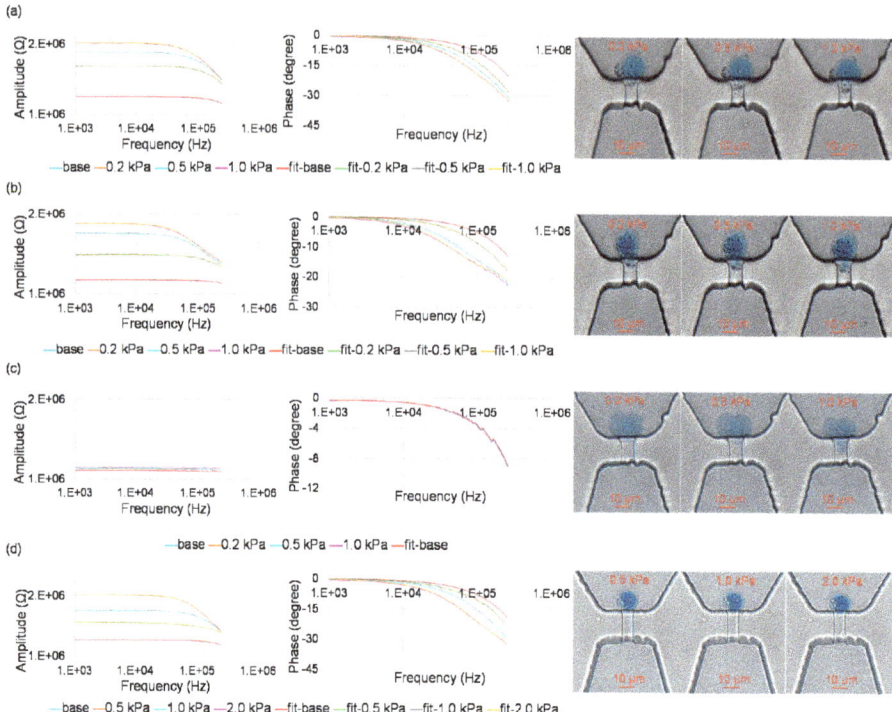

Figure 3. Measured impedance values with curve fitting and corresponding microscopic images for trapped single nuclei at the entrances of the constriction channels, under the conditions of IGEPAL® CA-630: 0.01%, constriction channel: 7 µm × 8 µm, aspiration pressure: 0.2 kPa, 0.5 kPa and 1.0 kPa, nucleus type: A549 (**a**), IGEPAL® CA-630: 0.02%, constriction channel: 7 µm × 8 µm, aspiration pressure: 0.2 kPa, 0.5 kPa and 1.0 kPa, nucleus type: A549 (**b**), IGEPAL® CA-630: 0.03%, constriction channel: 7 µm × 8 µm, aspiration pressure: 0.2 kPa, 0.5 kPa and 1.0 kPa, nucleus type: A549 (**c**), IGEPAL® CA-630: 0.01%, constriction channel: 5 µm × 5 µm, aspiration pressure: 0.5 kPa, 1.0 kPa and 2.0 kPa, nucleus type: SW620 (**d**).

As shown in Figure 3b, a similar trend was obtained for A549 nuclei under the conditions of 0.02% IGEPAL® CA-630 and the constriction-channel of 7 µm × 8 µm. As to A549 nuclei under the condition of 0.03% IGEPAL® CA-630, no significant differences of impedance with and without the trapping of single nucleus at the entrance of the constriction channel was noticed, indicating that the use of high concentration of IGEPAL® CA-630 can damage nuclear envelopes, which then cannot effectively block the electric lines in impedance measurements (see Figure 3c).

As shown in Figure 3d, a similar trend was obtained for SW620 nucleus under the conditions 0.01% IGEPAL® CA-630 and the constriction-channel of 5 µm × 5 µm. Curve fitting of measured impedance data for single nucleus was conducted where C_{ne}, R_{ne}, R_{np} and R_l were quantified to be 3.33 µF/cm^2, 1.83 Ω·cm^2, 0.20 MΩ and 0.65 MΩ (P_a = 0.5 kPa), 4.00 µF/cm^2, 0.62 Ω·cm^2, 0.20 MΩ and 1.75 MΩ (P_a = 1.0 kPa), 3.50 µF/cm^2, 2.64 Ω·cm^2, 0.40 MΩ and 1.80 MΩ (P_a = 2.0 kPa).

Figure 4 shows quantified single-nucleus electrical parameters of C_{ne}, R_{ne} and R_{np} as well as R_l. Under the conditions of 0.01% IGEPAL® CA-630 and the constriction-channel of 7 µm × 8 µm, C_{ne}, R_{ne}, R_{np} and R_l of the A549 nuclei were determined to be 4.02 ± 1.91 µF/cm^2, 1.67 ± 1.37 Ω·cm^2, 0.41 ± 0.21 MΩ and 0.43 ± 0.30 MΩ (N_n = 16, P_a = 0.2 kPa), 3.55 ± 1.66 µF/cm^2, 2.20 ± 1.52 Ω·cm^2, 0.43 ± 0.17 MΩ and 0.81 ± 0.57 MΩ (N_n = 16, P_a = 0.5 kPa), 3.94 ± 2.16 µF/cm^2, 2.82 ± 1.43 Ω·cm^2, 0.47 ± 0.17 MΩ and 1.49 ± 1.34 MΩ (N_n = 16, P_a = 1.0 kPa; see Figure 4a and Table 1).

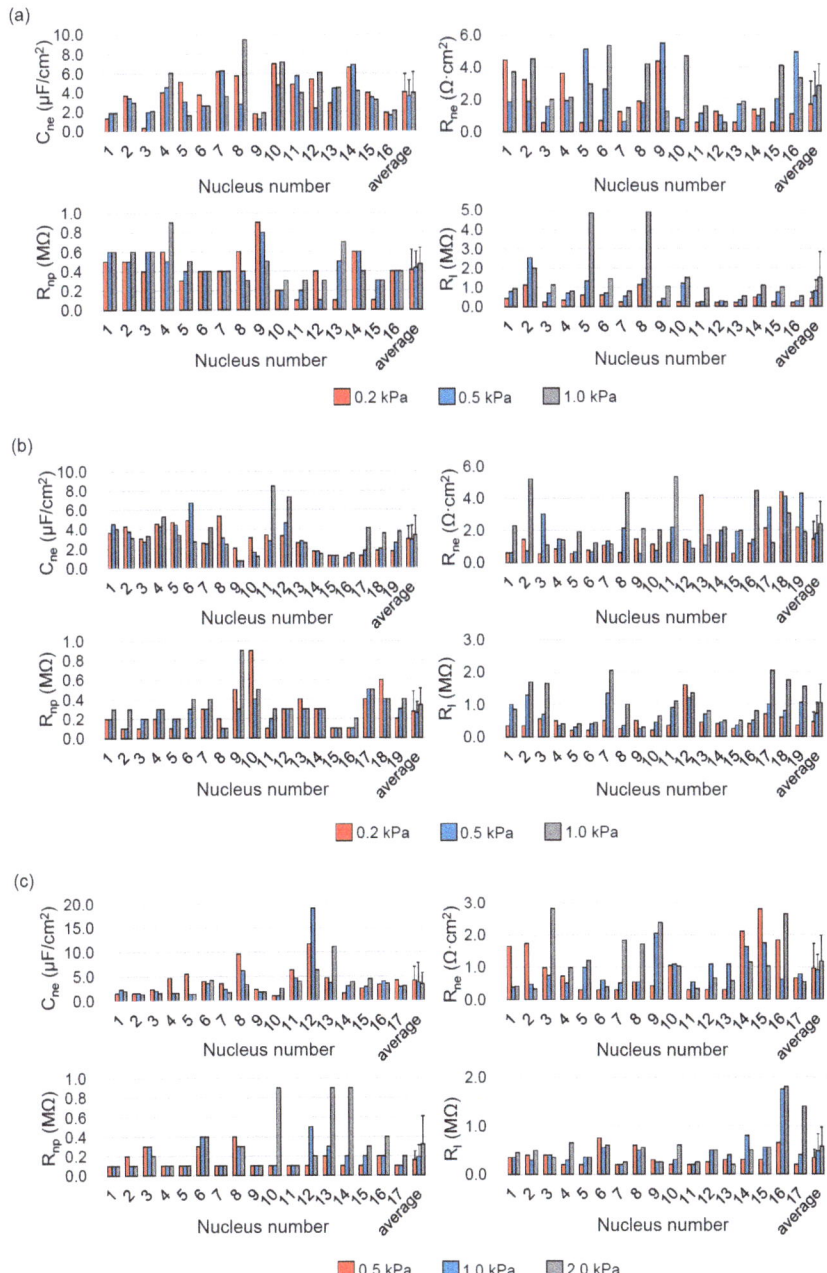

Figure 4. Quantified single-nucleus electrical parameters of C_{ne}, R_{ne} and R_{np} as well as R_l under the conditions of IGEPAL® CA-630: 0.01%, constriction channel: 7 μm × 8 μm, aspiration pressure: 0.2 kPa, 0.5 kPa and 1.0 kPa, nucleus type: A549, nucleus number: 16 (**a**), IGEPAL® CA-630: 0.02%, constriction channel: 7 μm × 8 μm, aspiration pressure: 0.2 kPa, 0.5 kPa and 1.0 kPa, nucleus type: A549, nucleus number: 19 (**b**), IGEPAL® CA-630: 0.01%, constriction channel: 5 μm × 5 μm, aspiration pressure: 0.5 kPa, 1.0 kPa and 2.0 kPa, nucleus type: SW620, nucleus number: 17 (**c**).

Table 1. A summary of quantified single-nucleus electrical parameters (e.g., C_{ne}, R_{ne} and R_{np}) using constriction channels under a variety of operation conditions.

Nucleus Type	Constriction Channel Dimensions	IGEPAL® CA-630	Aspiration Pressure (kPa)	C_{ne} ($\mu F/cm^2$)	R_{ne} ($\Omega \cdot cm^2$)	R_{np} ($M\Omega$)	Nucleus Number
A549	7 μm × 8 μm	0.01%	0.2	4.02 ± 1.91	1.67 ± 1.37	0.41 ± 0.21	N = 16
			0.5	3.55 ± 1.66	2.20 ± 1.52	0.43 ± 0.17	
			1.0	3.94 ± 2.16	2.82 ± 1.43	0.47 ± 0.17	
		0.02%	0.2	2.97 ± 1.32	1.45 ± 1.07	0.27 ± 0.21	N = 19
			0.5	2.90 ± 1.47	1.76 ± 1.14	0.26 ± 0.11	
			1.0	3.38 ± 1.96	2.38 ± 1.37	0.34 ± 0.17	
SW620	5 μm × 5 μm	0.01%	0.5	4.13 ± 2.82	0.96 ± 0.76	0.16 ± 0.09	N = 17
			1.0	3.74 ± 4.03	0.90 ± 0.48	0.19 ± 0.12	
			2.0	3.38 ± 2.37	1.18 ± 0.79	0.32 ± 0.29	

Under an arbitrary aspiration pressure, the obtained C_{ne} and R_{ne} were around 1 $\mu F/cm^2$ and 1 $\Omega \cdot cm^2$, which were consistent with previous studies [17,18,26,34,36,41]. With the increase of the aspiration pressure, R_l was observed to increase, suggesting the improvement in the sealing status between nuclear envelopes and inner walls of the constriction channel. Though the values of R_{ne} were noticed to increase when the aspiration pressure was increased from 0.2 kPa to 0.5 kPa and then 1.0 kPa, no differences with statistical significances for both C_{ne} and R_{ne} were located neighboring pressure values, indicating that the variations of aspiration pressure have no significantly side effects on the measurements of single-nucleus electrical properties.

Figure 4b shows quantified single-nucleus electrical parameters of C_{ne}, R_{ne} and R_{np} as well as R_l under the conditions of 0.02% IGEPAL® CA-630 and the constriction-channel of 7 μm × 8 μm. C_{ne} and R_{ne} of the A549 nuclei derived from 0.01% vs. 0.02% IGEPAL® CA-630 were compared as follows: 4.02 ± 1.91 $\mu F/cm^2$ vs. 2.97 ± 1.32 $\mu F/cm^2$ and 1.67 ± 1.37 $\Omega \cdot cm^2$ vs. 1.45 ± 1.07 $\Omega \cdot cm^2$ (P_a = 0.2 kPa), 3.55 ± 1.66 $\mu F/cm^2$ vs. 2.90 ± 1.47 $\mu F/cm^2$ and 2.20 ± 1.52 $\Omega \cdot cm^2$ vs. 1.76 ± 1.14 $\Omega \cdot cm^2$ (P_a = 0.5 kPa) and 3.94 ± 2.16 $\mu F/cm^2$ vs. 3.38 ± 1.96 $\mu F/cm^2$ and 2.82 ± 1.43 $\Omega \cdot cm^2$ vs. 2.38 ± 1.37 $\Omega \cdot cm^2$ (P_a = 1.0 kPa; see Figure 4b and Table 1). Though the values of C_{ne} and R_{ne} were observed to decrease when the percentage of IGEPAL® CA-630 was increased from 0.01% to 0.02%, no differences with statistical significances were located between these two groups, indicating that the use of IGEPAL® CA-630 for cell lysis at low concentrations has no significantly side effects on the measurements of single-nucleus electrical properties.

Figure 4c shows quantified single-nucleus electrical parameters of C_{ne}, R_{ne} and R_{np} as well as R_l under the conditions of 0.01% IGEPAL® CA-630 and the constriction-channel of 5 μm × 5 μm for SW620 nuclei. Under an arbitrary aspiration pressure, the obtained C_{ne} and R_{ne} were around 1 $\mu F/cm^2$ and 1 $\Omega \cdot cm^2$, which were consistent with previous studies [17,18,26,34,36,41]. Again, no differences with statistical significances were located among values of C_{ne} and R_{ne} under three different aspiration pressures, confirming that the variations of aspiration pressure have no significantly side effects on electrical properties of nuclear envelopes. Compared with the values of C_{ne} and R_{ne} obtained from A549 nuclei, the obtained values of SW620 nuclei were lower in both C_{ne} (4.00 ± 3.03 $\mu F/cm^2$ vs. 3.75 ± 3.17 $\mu F/cm^2$) and R_{ne} (1.69 ± 1.29 $\Omega \cdot cm^2$ vs. 1.01 ± 0.70 $\Omega \cdot cm^2$). Furthermore, significant differences with statistics were only located in R_{ne}, suggesting that nuclei of A549 and SW620 are different in equivalent resistance of nuclear envelopes while they have comparable capacitive properties.

4. Conclusions and Future Developments

This study demonstrated the feasibility of the microfluidic device by quantifying electrical properties of A549 nuclei, and reporting electrical properties of single nuclei independent from key parameters in nuclear isolation and measurement. Then this device was also used to characterize electrical properties of SW620 nuclei, which indicated that this approach could be used to test multiple nucleus types.

Future technical developments may aspirate single nuclei rapidly through the microfluidic constriction channel, enabling high-throughput quantification of single-nucleus electrical properties. Then the approach can be further expanded to high-throughput quantify multiple electrical parameters of cell membrane and nuclear portions, enabling cell type classification and cell status evaluation in a label-free manner.

Supplementary Materials: The following are available online at http://www.mdpi.com/2072-666X/10/11/740/s1. Table S1: A summary of previously reported electrical parameters of single.

Author Contributions: H.L., J.W. and J.C. designed experiments; H.L., Y.Z. and H.T. contributed to cell treatments, device fabrication and device operation; H.L., Y.Z. and J.C. contributed to data processing; Y.Z., D.C. and J.C. contributed to modeling development; and H.L., J.W. and J.C. drafted the manuscript.

Funding: This research was funded by the National Natural Science Foundation of China (Grant No. 61431019, 61825107, 61671430, 61922079); Key Project (QYZDB-SSW-JSC011), Instrument Development Program, Youth Innovation Promotion Association and Interdisciplinary Innovation Team of Chinese Academy of Sciences; Instrument Development of Beijing Municipal Science & Technology Commission (Z181100009518001).

Acknowledgments: The authors would like to acknowledge helpful discussions with Ke Wang (Beijing University of Posts and Telecommunications) about the extraction of single nuclei.

Conflicts of Interest: The authors declare no conflicts of interest.

References

1. Loewenstein, W.R.; Kanno, Y. Some electrical properties of the membrane of a cell nucleus. *Nature* **1962**, *195*, 462–464. [CrossRef]
2. Loewenstein, W.R.; Kanno, Y. Some electrical properties of a nuclear membrane examined with a microelectrode. *J. Gen Physiol.* **1963**, *46*, 1123–1140. [CrossRef] [PubMed]
3. Loewenstein, W.R.; Kanno, Y. The electrical conductance and potential across the membrane of some cell nuclei. *J. Cell Biol.* **1963**, *16*, 421–425. [CrossRef] [PubMed]
4. Loewenstein, W.R. Permeability of a nuclear membrane: changes during normal development and changes induced by growth hormone. *Science* **1965**, *150*, 909–910. [PubMed]
5. Loewenstein, W.R.; Kanno, Y.; Ito, S. Permeability of nuclear membranes. *Ann. N. Y. Acad. Sci.* **1966**, *137*, 708–716. [CrossRef]
6. Palmer, L.G.; Civan, M.M. Distribution of Na $^+$, K $^+$ and Cl$^-$ between nucleus and cytoplasm in chironomus salivary gland cells. *J. Membr. Biol.* **1977**, *33*, 41–61. [CrossRef]
7. Kanno, Y.; Loewenstein, W.R. A study of the nucleus and cell membranes of oocytes with an intra-cellular electrode. *Exp. Cell Res.* **1963**, *31*, 149–166. [CrossRef]
8. Kanno, Y.; Ashman, R.F.; Loewenstein, W.R. Nucleus and cell membrane conductance in marine oocytes. *Exp. Cell Res.* **1965**, *39*, 184–189. [CrossRef]
9. Dale, B. Voltage clamp of the nuclear envelope. *Proc. R. Soc. B* **1994**, *255*, 119–124.
10. Danker, T. Using atomic force microscopy to investigate patch-clamped nuclear membrane. *Cell Biol. Int.* **1997**, *21*, 747–757. [CrossRef]
11. Mazzanti, M. Ion permeability of the nuclear envelope. *News Physiol. Sci.* **1998**, *13*, 44–50. [CrossRef] [PubMed]
12. Danker, T. Nuclear hourglass technique: An approach that detects electrically open nuclear pores in xenopus laevis oocyte. *Proc. Natl. Acad. Sci. USA* **1999**, *96*, 13530–13535. [CrossRef] [PubMed]
13. Danker, T. Electrophoretic plugging of nuclear pores by using the nuclear hourglass technique. *J. Membr. Biol.* **2001**, *184*, 91–99. [CrossRef] [PubMed]
14. Ziervogel, H. Electrorotation of lymphocytes–the influence of membrane events and nucleus. *Biosci. Rep.* **1986**, *6*, 973–982. [CrossRef] [PubMed]
15. Asami, K.; Takahashi, Y.; Takashima, S. Dielectric properties of mouse lymphocytes and erythrocytes. *Biochim. Biophys. Acta* **1989**, *1010*, 49–55. [CrossRef]
16. Griffith, A.W.; Cooper, J.M. Single-cell measurements of human neutrophil activation using electrorotation. *Anal. Chem.* **1998**, *70*, 2607–2612. [CrossRef]
17. Polevaya, Y. Time domain dielectric spectroscopy study of human cells. II. normal and malignant white blood cells. *Biochim. Biophys. Acta* **1999**, *1419*, 257–271. [CrossRef]

18. Ermolina, I.; Polevaya, Y.; Feldman, Y. Study of normal and malignant white blood cells by time domain dielectric spectroscopy. *IEEE Trans. Dielectr. Electr. Insul.* **2001**, *8*, 253–261. [CrossRef]
19. Egger, M.; Donath, E. Electrorotation measurements of diamide-induced platelet activation changes. *Biophys. J.* **1995**, *68*, 364. [CrossRef]
20. Feldman, Y.; Ermolina, I.; Hayashi, Y. Time domain dielectric spectroscopy study of biological systems. *IEEE Trans. Dielectr. Electr. Insul.* **2003**, *10*, 728–753. [CrossRef]
21. Raicu, V.; Raicu, G.; Turcu, G. Dielectric properties of yeast cells as simulated by the two-shell model. *Biochim. Biophys. Acta* **1996**, *1274*, 143–148. [CrossRef]
22. Matzke, A.J.; Weiger, T.M.; Matzke, M. Ion channels at the nucleus: electrophysiology meets the genome. *Mol. Plant* **2010**, *3*, 642–652. [CrossRef] [PubMed]
23. Giulian, D.; Diacumakos, E.G. Electrophysiological mapping of compartments within a mammalian-cell. *J. Cell Biol.* **1977**, *72*, 86–103. [CrossRef] [PubMed]
24. Irimajiri, A. Passive electrical properties of cultured murine lymphoblast (L 5178 Y) with reference to its cytoplasmic membrane, nuclear envelope, and intracellular phases. *J. Membr. Biol.* **1978**, *38*, 209–232. [CrossRef] [PubMed]
25. Gimsa, J. Dielectrophoresis and electrorotation of neurospora slime and murine myeloma cells. *Biophys. J.* **1991**, *60*, 749–760. [CrossRef]
26. Oberleithner, H. Imaging nuclear pores of aldosterone-sensitive kidney cells by atomic force microscopy. *Proc. Natl. Acad. Sci. USA* **1994**, *91*, 9784–9788. [CrossRef]
27. Garner, A.L. Time domain dielectric spectroscopy measurements of HL-60 cell suspensions after microsecond and nanosecond electrical pulses. *IEEE Trans. Plasma Sci.* **2004**, *32*, 2073–2084. [CrossRef]
28. Garner, A.L. Ultrashort electric pulse induced changes in cellular dielectric properties. *Biochem. Biophys. Res. Commun.* **2007**, *362*, 139–144. [CrossRef]
29. Massimi, M. Dielectric Characterization of hepatocytes in suspension and embedded into two different polymeric scaffolds. *Colloids Surf. B Biointerfaces* **2013**, *102*, 700–707. [CrossRef]
30. Moisescu, M.G. Changes of cell electrical parameters induced by electroporation. A dielectrophoresis study. *Biochim. Biophys. Acta* **2013**, *1828*, 365–372. [CrossRef]
31. Zhuang, J. Pulsed electric field induced dielectric evolution of mammalian cells. In Proceedings of the IEEE International Power Modulator and High Voltage Conference, San Diego, CA, USA, 3–7 June 2013.
32. Anderson, S.E.; Bau, H.H. Electrical detection of cellular penetration during microinjection with carbon nanopipettes. *Nanotechnology* **2014**, *25*, 245102. [CrossRef]
33. Sano, M.B. In-vitro bipolar nano- and microsecond electro-pulse bursts for irreversible electroporation therapies. *Bioelectrochemistry* **2014**, *100*, 69–79. [CrossRef] [PubMed]
34. Stacey, M.W.; Sabuncu, A.C.; Beskok, A. Dielectric characterization of costal cartilage chondrocytes. *Biochim. Biophys. Acta* **2014**, *1840*, 146–152. [CrossRef] [PubMed]
35. Murovec, T. Modeling of transmembrane potential in realistic multicellular structures before electroporation. *Bioelectrochem. Bioenerg.* **2016**, *111*, 2286–2295. [CrossRef] [PubMed]
36. Sabuncu, A.C. Microfluidic impedance spectroscopy as a tool for quantitative biology and biotechnology. *Biomicrofluidics* **2012**, *6*, 137–152. [CrossRef]
37. Zhou, Y. Single cell studies of mouse embryonic stem cell (mESC) differentiation by electrical impedance measurements in a microfluidic device. *Biosens. Bioelectron.* **2016**, *81*, 249–258. [CrossRef]
38. Reynolds, C.R.; Tedeschi, H. Permeability properties of mammalian cell nuclei in living cells and in vitro. *J. Cell Sci.* **1984**, *70*, 197–207.
39. Mazzanti, M.; DeFeliceI, L.J. Ion channel in the nuclear envelope. *Nature* **1990**, *343*, 764–767. [CrossRef]
40. Matzke, A.J.M. A large conductance ion channel in the nuclear envelope of higher plant cell. *FEBS Lett.* **1992**, *302*, 81–85. [CrossRef]
41. Raicu, V. Dielectric properties of rat liver in vivo: Analysis by modeling hepatocytes in the tissue architecture. *Bioelectrochem. Bioenerg.* **1998**, *47*, 333–342. [CrossRef]
42. Vanderstraeten, J.; Vander Vorst, A. Theoretical evaluation of dielectric absorption of microwave energy at the scale of nucleic acids. *Bioelectromagnetics* **2004**, *25*, 380–389. [CrossRef] [PubMed]
43. Bai, W.; Zhao, K.S.; Asami, K. Dielectric properties of E. coli cell as simulated by the three-shell spheroidal model. *Bioelectrochem. Bioenerg.* **2006**, *122*, 136–142. [CrossRef] [PubMed]

44. Gerber, H.L.; Bassi, A.; Tseng, C.C. Determining a pulse coupling to subcellular components with a frequency-domain transfer function. *IEEE Trans. Plasma Sci.* **2006**, *34*, 1425–1430. [CrossRef]
45. Kotnik, T.; Miklavc, D. Theoretical evaluation of voltage inducement on internal membranes of biological cells exposed to electric fields. *Biophys. J.* **2006**, *90*, 480–491. [CrossRef] [PubMed]
46. Shibatani, S. Electrorotation of vacuoles isolated from barley mesophyll cells. *Bioelectrochem. Bioenerg.* **1993**, *29*, 327–335. [CrossRef]
47. Chun-Chieh, C.; Ke, W.; Yi, Z.; Deyong, C.; Beiyuan, F.; Chia-Hsun, H. Mechanical property characterization of hundreds of single nuclei based on microfluidic constriction channel. *Cytom. A* **2018**, *93*, 1–7.
48. Gravesen, P.; Branebjerg, J.; Jensen, O.S. Microfluidics-a review. *J. Micromech. Microeng.* **1993**, *3*, 168. [CrossRef]
49. Gross, A.; Schoendube, J.; Zimmermann, S.; Steeb, M.; Zengerle, R.; Koltay, P. Technologies for single-cell isolation. *Int. J. Mol. Sci.* **2015**, *16*, 16897–16919. [CrossRef]
50. Shields, C.W., IV; Reyes, C.D.; López, G.P. Microfluidic cell sorting: a review of the advances in the separation of cells from debulking to rare cell isolation. *Lab Chip* **2015**, *15*, 1230–1249.
51. Warkiani, M.E.; Wu, L.; Tay, A.K.P.; Han, J. Large-volume microfluidic cell sorting for biomedical applications. *Annu. Rev. Biomed. Eng.* **2015**, *17*, 1. [CrossRef]
52. Mu, X.; Zheng, W.; Sun, J.; Zhang, W.; Jiang, X. Microfluidics for manipulating cells. *Small* **2013**, *9*, 9–21. [CrossRef] [PubMed]
53. Ayliffe, H.E.; Frazier, A.B.; Rabbitt, R.D. Electric impedance spectroscopy using microchannels with integrated metal electrodes. *IEEE J. Microelectromech. Syst.* **1999**, *8*, 50–57. [CrossRef]

© 2019 by the authors. Licensee MDPI, Basel, Switzerland. This article is an open access article distributed under the terms and conditions of the Creative Commons Attribution (CC BY) license (http://creativecommons.org/licenses/by/4.0/).

Article

Design and Clinical Application of an Integrated Microfluidic Device for Circulating Tumor Cells Isolation and Single-Cell Analysis

Mingxin Xu [1,†], Wenwen Liu [2,†], Kun Zou [3,†], Song Wei [1], Xinri Zhang [4,*], Encheng Li [1,*] and Qi Wang [1,*]

1. Department of Respiratory Medicine, The Second Hospital Affiliated to Dalian Medical University, No. 467 Zhongshan Road, Dalian 116023, China; xumingxin5426@163.com (M.X.); 18842637303@163.com (S.W.)
2. Cancer Translational Medicine Research Center, The Second Hospital Affiliated to Dalian Medical University, No. 467 Zhongshan Road, Dalian 116023, China; liuwenwenphd@163.com
3. Department of Radiation Oncology, The First Hospital Affiliated to Dalian Medical University, No. 222 Zhongshan Road, Dalian 116011, China; zoukun29@163.com
4. Department of Respiratory and Critical Care Medicine, The First Hospital, Shanxi Medical University, No. 85 Jiefang South Road, Taiyuan 030001, China
* Correspondence: ykdzxr61@163.com (X.Z.); doctorliencheng@163.com (E.L.); wqdlmu@dmu.edu.cn (Q.W.)
† These authors contribute to this manuscript equally.

Citation: Xu, M.; Liu, W.; Zou, K.; Wei, S.; Zhang, X.; Li, E.; Wang, Q. Design and Clinical Application of an Integrated Microfluidic Device for Circulating Tumor Cells Isolation and Single-Cell Analysis. *Micromachines* 2021, *12*, 49. https://doi.org/10.3390/mi12010049

Received: 25 November 2020
Accepted: 30 December 2020
Published: 2 January 2021

Publisher's Note: MDPI stays neutral with regard to jurisdictional claims in published maps and institutional affiliations.

Copyright: © 2021 by the authors. Licensee MDPI, Basel, Switzerland. This article is an open access article distributed under the terms and conditions of the Creative Commons Attribution (CC BY) license (https://creativecommons.org/licenses/by/4.0/).

Abstract: Circulating tumor cells (CTCs) have been considered as an alternative to tissue biopsy for providing both germline-specific and tumor-derived genetic variations. Single-cell analysis of CTCs enables in-depth investigation of tumor heterogeneity and individualized clinical assessment. However, common CTC enrichment techniques generally have limitations of low throughput and cell damage. Herein, based on micropore-arrayed filtration membrane and microfluidic chip, we established an integrated CTC isolation platform with high-throughput, high-efficiency, and less cell damage. We observed a capture rate of around 85% and a purity of 60.4% by spiking tumor cells (PC-9) into healthy blood samples. Detection of CTCs from lung cancer patients demonstrated a positive detectable rate of 87.5%. Additionally, single CTCs, ctDNA and liver biopsy tissue of a representative advanced lung cancer patient were collected and sequenced, which revealed comprehensive genetic information of CTCs while reflected the differences in genetic profiles between different biological samples. This work provides a promising tool for CTCs isolation and further analysis at single-cell resolution with potential clinical value.

Keywords: CTC-isolation; microfluidics; single-cell analysis; high-throughput sequencing; lung cancer

1. Introduction

Cancer is a major medical problem endangering human health, while metastasis remains the most common cause of tumor-related death. According to the "seed and soil" hypothesis proposed by Paget in 1889 [1], metastasis is a complex process when rare metastatic "seeds" shed from primary/secondary tumor lesions into the blood circulation and seek suitable "soil" to colonize. These "seeds", also called circulating tumor cells (CTCs), are considered not only as culprits of metastasis but a promising alternative to tissue biopsy [2]. While enumeration of CTCs has been seen as a prognostic indicator in many cancers, genomic analysis is the prime way to elucidate oncogenic profiles useful for tumor characterization and personalized treatment [3].

In recent years, the rapid development of microfluidics has brought lots of new findings in cellular biology. For cell biomechanics research, the application of microfluidic technology has allowed us to characterize the deformability of live cells, even their nucleuses, to physical stress [4,5]. For drug-related research, with microfluidics, a new approach has been developed and demonstrated for studying the physiological reactions of viable but non-culturable (VBNC) cells to external disturbances, like drug treatment, which may unravel the mechanism of antibiotic resistance [6]. While different procedures have been described to quantify the intracellular accumulation and investigate the membrane-associated transportation of given antibiotics [7]. Similarly, extensive works have been carried out to achieve drug mix and antibiotic diffusion characterization on chip [8,9]. In addition to all the above, microfluidic techniques have also promoted studies of tumor, especially CTCs greatly. Advancements in microfluidic technology have contributed to building numerous new devices for CTCs studies, from detection to cell culture [10]. More importantly, progression in high-throughput sequencing—especially single-cell sequencing—has enabled genomics and transcriptomics analysis of CTCs at single-cell resolution [11]. Single-cell analysis has been broadly used in describing oncogenic mutation patterns, discovering tumor heterogeneities, even guiding treatment options [12–14]. Given the emergence of single-cell analysis, it is imperative to establish CTC isolation platforms that are compatible with subsequent genomic analysis. However, common CTC enrichment methods, such as density gradient separation and CellSearch, often have limitations including cell loss, cell damage and time-consuming [15,16]. Therefore, to obtain CTCs qualified for single-cell sequencing, there is an urgent need for a robust device that can achieve high-throughput, high-efficiency CTC isolation while minimizing cell damage.

In this study, we report an integrated device based on micropore-arrayed filtration membrane and microfluidic chip to isolate CTCs for single-cell analysis. The performance of this device has been confirmed on the cultured cell line and patient blood samples of lung cancer. Additionally, to further verify the clinical application of this platform, single-cell sequencing was performed on single CTCs from a representative advanced lung cancer patient.

2. Materials and Methods

2.1. Microfluidic Chips Fabrication, Filtration Membranes Preparation and System Package

This device consisted of two parts. The micropore-arrayed filtration membranes we used in this study were made of Parylene C by a molding technique. The finished membranes were packaged with Teflon holders, then fixed with magnetic rings before the filtration operation. Details about this part could be found in our previous paper [17].

The magnetic purifying device included a microfluidic chip, a permanent magnet and an electronic control system. According to our previous work [18], the chip was made of polymethyl methacrylate (PMMA) (Sigma-Aldrich, Darmstadt, Germany), and a Laser Engraving Machine (LS100CO2, Gravograph, Shanghai, China) was used for creating microchannels on it. The chip consisted of three layers, which were pressed together for 10 min at 120 °C by a Thermocompressor (CARVER, Muscatine, IA, USA). The magnet was placed under the chip, whose movement could be regulated by an electronic system controlled by a C Language-based software.

2.2. CTC Detection

Healthy peripheral blood samples spiked with GFP-expressed PC-9 cells (PC9-GFP) or peripheral blood samples from lung cancer patients (2 mL) were loaded into this device. After filtration, the packaged membrane structure was quickly reversed and 500 μL PBS was added from the upper opening to wash captured blood cells off the membrane, then cells were collected into a tube placed directly below the outlet of the filtration structure along with the fluid. Next, CTCs and residual white blood cells (WBCs) were injected into the purifying chip through the inlet, while CD45 dynabeads were added into it through the outlet. Under the control of the electronic system, dynabeads were forced to run around the reaction chamber of the chip via magnet to bind WBCs, and the best binding time was 20 min. When the reaction was over, the magnet was allowed to stop at the bottom of the reaction chamber to attract the dynabeads, then the supernatant with unbounded WBCs and CTCs was transferred into a culture dish for further identification.

2.3. Single-Cell Isolation

A micromanipulator was used for separating single CTCs as described previously [19]. Briefly, for manual cell picking, a tapered ultrathin glass capillary (Friends Honesty Life Sciences Company, Beijing, China) was connected with a Teflon tube to aspirate and dispense single cells, under the negative and positive pressures provided by two constant-pressure pumps (WH-PMPP-12, Wenhao Co., Suzhou, China). The aspirated liquid containing the selected single cells was transferred to a centrifuge tube with lysis buffer for DNA extraction and high-throughput sequencing.

2.4. CTCs Identification

The captured CTCs were placed into a cell culture dish with phosphate buffer saline (PBS) for immunofluorescence staining. The CTCs were identified using Alexa Fluor 594 conjugated cytokeratin (Pan-reactive) antibody (1:100, Novus Biologicals, Littleton, CO, USA). Alexa Flour 488 mouse anti-human CD45 (1:100, Abcam, Cambridge, UK) antibody was used to differentiate white blood cells (WBCs). The nuclei were stained with Hoechst (1:2000, Life Technologies, Waltham, MA, USA). Cells were observed under an inverted fluorescence microscope (DMI3000B; Leica Microsystems, Buffalo Grove, IL, USA). CK+/DAPI+/CD45− phenotype was considered to be CTCs, while CD45+/DAPI+/CK− phenotype confirmed WBCs.

2.5. Cell Line Culture and Preparation

Human lung adenocarcinoma cell line PC-9, obtained from the American Type Culture Collection (ATCC), was stably transfected with a green fluorescent protein (GFP) before used as a model to mimic actual CTCs in efficacy evaluation of this developed device. Cells were cultured in 10% fetal bovine serum and 1% penicillin/streptomycin added Dulbecco's Modified Eagle Medium (DMEM)/high-glucose medium (Gibco, Carlsbad, CA, USA; Invitrogen, Carlsbad, CA, USA), with 5% CO_2 at 37 °C. Cells were incubated with 0.05% trypsin–ethylenediaminetetraacetic acid (EDTA) at 37 °C for 3 min, then suspended and diluted to the desired concentration.

2.6. Human Peripheral Blood Samples

The blood samples from lung cancer patients and healthy volunteers were obtained from The Second Affiliated Hospital of Dalian Medical University, People's Republic of China. All subjects were informed consent. 2 mL peripheral blood of each subject was collected into EDTA containing tubes and processed within 6 h. The clinicopathological characteristics of patients were recorded as well. This study was approved by the Ethics Review Committee of the Second Affiliated Hospital of Dalian Medical University (2018-048). All experiments were conducted according to ethical and safe research practices consisting of human subjects or blood and the Helsinki Declaration of 1975, as revised in 2013.

2.7. DNA Sequencing and Bioinformatics Analysis

DNA sequencing was conducted by Haplox (Shenzhen, China). Whole-exome sequencing was performed on liver biopsy tissue, while HapOnco-605 panel sequencing was performed on single CTC and ctDNA samples. The generated library was sequenced on Illumina NovaSeq6000 or HiSeq X platform (Illumina, San Diego, CA, USA), according to PE150 strategies.

The sequencing data was processed by HPS Gene Technology Co., Ltd. (Tianjin, China). FastQC was used to achieve quality control and filter low-quality raw data. The mapping to the human reference genome hg19 was performed with a Burrows-Wheeler Aligner. Somatic SNVs and InDels calling were conducted with muTect and Strelka respectively, while annotation with ANNOVAR. The number of identified SNVs and InDels, together with the proportion of allelic mutations were graphed using R language (ggplot2). R language (maftools) was used to generate the list of genes whose mutation rates ranking top 10. For comparative analysis, R language (VennDiagram) was used to show the number of common or unique mutations/genes among different samples.

2.8. Statistical Analysis

All statistical analyses were conducted by GraphPad Prism8. Data were expressed as mean ± standard error of mean (SEM).

3. Results

3.1. Design and Performance Verification of the Integrated Microfluidic Device

The newly established integrated device was shown in Figure 1. The device consisted of two parts, a filtration system on the top and a magnetic microfluidic chip below. The filtration system included a rotatable packaged micropore-arrayed membrane-based structure for CTC isolation, a removable connecting pipe and a centrifuge tube for the storage of waste or samples remained to be purified. When considered as a whole, this part was about 125 mm high, with the maximum diameter of 40 mm. The position of all components above could be adjusted as needed, and the bottom of the connecting pipe could be placed into the centrifuge tube to prevent liquid splashing. The area of the membrane is 20×20 mm, while the diameter of the micropore is 10 μm. As reported in our previous work, the throughput of this membrane could reach up to 17 mL/min for undiluted whole blood, driven by gravity. During filtration, all red blood cells and platelets, as well as most WBCs could pass through the membrane and flow into the waste collection tube. Since the size distributions of CTCs (12–25 μm) and WBCs (5–20 μm) partially overlap [20], CTCs and some WBCs with larger size would be captured on the membrane. The capture rate for lung cancer cells spiked in 5 mL unprocessed whole blood was $83.2 \pm 6.2\%$, with the number of WBCs decreased from 10^9/mL to $10^4 \sim 10^5$/mL without clogging. After filtration, CTCs and residual WBCs were rinsed off the membrane for further purification. To this end, an automatic magnetic microfluidic chip for negative isolation was employed as described before [18], whose volume was enough to hold 500 μL flushing fluid obtained from the first step. The overall height of the chip and the magnetic base was about 70 mm, and the maximum diameter of the base was 50 mm. The best mixing ratio and binding time for this part had been validated as 10:1 (dynabeads/WBCs) and 20 min in our previous experiments [18]. All parts of this device were fixed on a shelf, whose height and width were 250 and 80 mm respectively.

To test the performance of our integrated device, we spiked PC9-GFP cells into healthy blood samples at concentrations of 10^1 to 10^4 cells/mL to mimic actual CTCs (Figure 2a,b). The overall PC9-GFP cells capture rate of this device was calculated as (the number of captured CTCs/the number of total CTCs) $\times$ 100%, which was 84.0%, 84.8%, 85.4% and 85.9% for different concentrations (Figure 2c and Supplementary Materials Table S1). As for capture purity, which was calculated as (the number of captured CTCs/the number of total isolated nucleated cells) $\times$ 100%, we found about 60.4% at the concentration of 10^4 cells/mL (Table S2).

3.2. Detection of CTCs from Blood Samples of Lung Cancer Patients

For clinical validation of this device, CTC detection was conducted on peripheral blood samples from patients. 16 lung cancer patients of different stages and 4 non-tumor patients as controls, were enrolled in this study. 2 mL blood sample was drawn from each subject, and their clinicopathological characteristics, together with CTC detection results were recorded in Table 1. Captured CTCs were identified by Immunofluorescence staining of CK and CD45. Cells with a phenotype of CK+/CD45−/DAPI+ were recognized as CTCs, while those with CK−/CD45+/DAPI+ were recognized as WBCs (Figure 3a). 14 of 16 lung cancer patients showed positive results of CTC detection (87.5% detectable rate). As shown in Figure 3b,c, the counts of CTCs varied between patients in different stages, while no CTC was observed in any non-tumor patient.

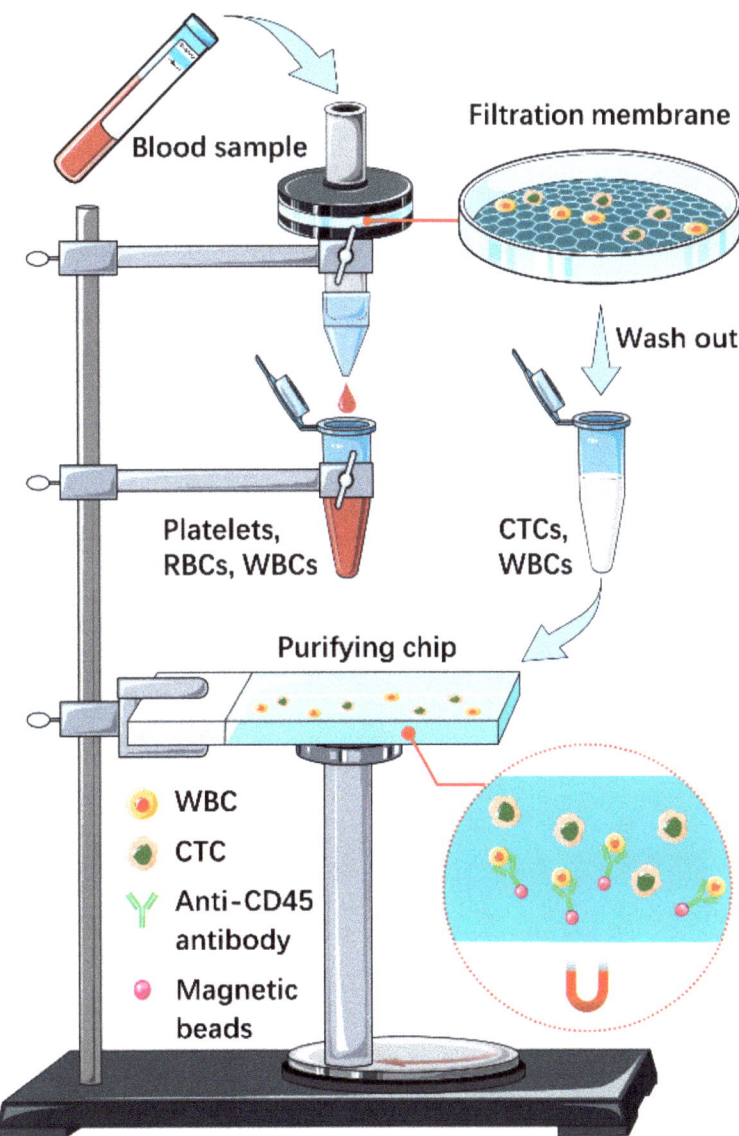

Figure 1. Schematic illustration of the integrated microfluidic circulating tumor cells (CTC) isolation platform. Taking lung cancer as an example, 2 mL blood sample was loaded into the device. All red blood cells (RBCs), platelets and most WBCs were filtered, while CTCs and some WBCs of large size were captured. Then CTCs and residual WBCs were washed off the membrane and transferred into the purifying chip, where WBCs would be further removed by CD45 dynabeads. Finally, the supernatant with CTCs and free WBCs was collected for identification and single-cell isolation. This work is licensed under a Creative Commons Attribution 3.0 Unported License. It is attributed to Mingxin Xu.

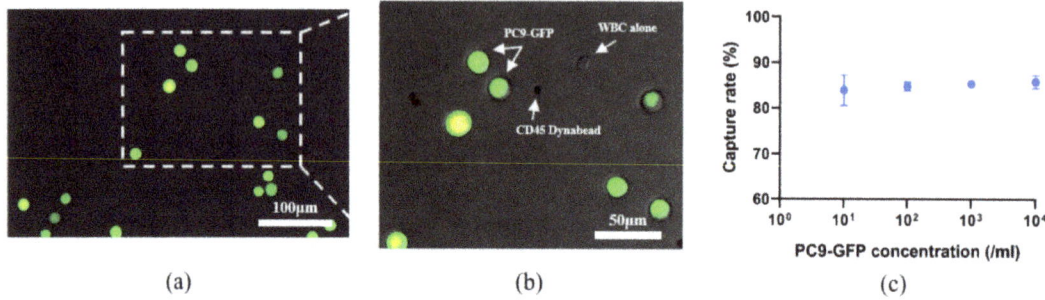

Figure 2. Capture of PC9-GFP cells by the integrated device. (**a**,**b**) Representative merged fluorescent field and bright images of PC9-GFP cells, white blood cells (WBCs) and CD45 dynabeads (as indicated by the arrows) after processing by the novel device (**a**) (magnification, ×200), (**b**) (magnification, ×400). (**c**) Capture rate of the integrated device at different concentrations of PC9-GFP cells. Data were expressed as mean ± standard error of mean (SEM).

Table 1. Clinical characteristics of 20 patients.

No.	Gender	Diagnosis	Stage	Histological Type	CTC Count (N)
1	Male	Lung cancer	IV	Adenocarcinoma	43
2	Female	Lung cancer	IV	Adenocarcinoma	21
3	Male	Lung cancer	IV	Small cell carcinoma	6
4	Male	Lung cancer	IV	Large cell carcinoma	10
5	Female	Lung cancer	IV	Adenocarcinoma	31
6	Male	Lung cancer	IV	Adenocarcinoma	9
7	Male	Lung cancer	IV	Adenocarcinoma	0
8	Female	Lung cancer	IIIC	Adenocarcinoma	15
9	Female	Lung cancer	IV	Adenocarcinoma	13
10	Male	Lung cancer	Postoperative	Adenocarcinoma	0
11	Male	Lung cancer	IIA	Adenocarcinoma	14
12	Female	Lung cancer	IV	Adenocarcinoma	9
13	Female	Lung cancer	IV	Adenocarcinoma	12
14	Male	Lung cancer	IV	Squamous cell carcinoma	17
15	Female	Lung cancer	IIIB	Adenocarcinoma	6
16	Male	Lung cancer	IA	Adenocarcinoma	26
17	Male	Pneumonia	—[2]	—[2]	0
18	Male	COPD [1]	—[2]	—[2]	0
19	Male	COPD [1]	—[2]	—[2]	0
20	Male	COPD [1]	—[2]	—[2]	0

[1] COPD, chronic obstructive pulmonary disease; [2] "—", no such information.

3.3. Single-Cell Analysis of CTC from a Representative Advanced Lung Cancer Patient

To further explore the underlying heterogeneity of lung cancer at single-cell resolution, high-throughput sequencing was performed on circulating tumor DNA (ctDNA), single CTCs and liver metastases from a representative advanced lung cancer patient (patient #1).

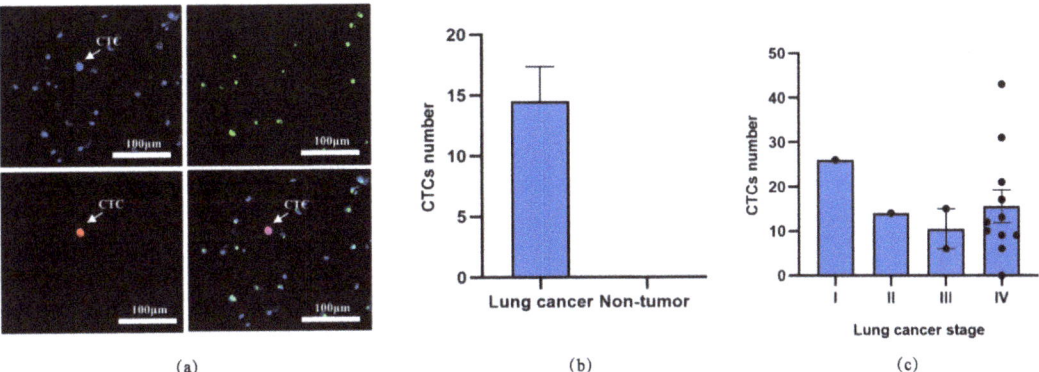

Figure 3. Analysis of captured CTCs from blood samples of patients. (**a**) Representative fluorescent staining images of CTC detected in a blood sample of a lung cancer patient. The blue color referred to nuclear staining (DAPI), the green color referred to CD45, and the red color demonstrated CK expression. Magnification, ×200. (**b**) The average number of CTCs in lung cancer patients (n = 16) and non-tumor patients (n = 4). (**c**) The average number of CTCs in lung cancer patients of different stages (Stage I, n = 1; Stage II, n = 1; Stage III, n = 2; Stage IV, n = 11).

As shown in Figure 4, patient #1 was a 42-year-old male who was admitted to the hospital due to cough and blood in the sputum. The findings of computed tomography (CT) imaging and positron emission tomography (PET)-CT both referred to a suspicious malignant nodule in the lower lobe of the right lung, with multiple lesions in the lung, liver and spine. At this time, 34 CTCs were found in his blood sample through our integrated device (2 mL). Several days later, according to the pathological analysis of bronchoscopy and needle biopsies, he was diagnosed with non-small cell lung cancer (NSCLC) with multiple metastases in the liver, bone, and lung (Lung adenocarcinoma, Stage IV). After diagnosis, another CTC detection was performed and 43 CTCs were found in a 1 mL blood sample. Single CTCs were isolated, while ctDNA and the remaining liver biopsy tissue were collected at the same time. Mutation analysis of metastatic liver tissue showed epidermal growth factor receptor (EGFR) mutation, so this patient was given EGFR-tyrosine kinase inhibitor (EGFR-TKI) Tarceva (Erlotinib). After one cycle, significant tumor regression was observed and only 1 CTC was found in a 1 mL blood sample. In the next year, this patient completed 3 sessions of radiotherapy with concurrent administration of Tarceva/Anlotinib and eventually died of sudden breathing difficulties. 4 single CTCs (3 from sample before treatment and 1 from sample after treatment) together with ctDNA and liver biopsy tissue of this patient was sent for DNA sequencing as summarized in Table 2, while his lymphocytes were used as a control to recognize germline-specific genetic variation.

We got sequencing results of 4 samples because the single CTCs before treatment (S090) was contaminated. Through pairwise analysis, we identified 1330 single nucleotide variations (SNVs) and 186 insertion-deletions (InDels) in total. The number of SNVs and InDels of each sample was shown in Table 2, while their distribution on the genome was analyzed in Figure 5a,b. The single CTC sample after treatment (S094) had the highest count of SNVs and InDels, which was mainly distributed in exonic and intronic regions, followed by the tissue sample (S145). As demonstrated in Figure 5c, the base mutation profiles of tissue and single CTC sample were similar, while more complex than that of ctDNA samples, especially the ctDNA sample after treatment (S63). The top 10 somatic mutations among 4 samples were listed in Figure 5d. For tissue/ctDNA samples before treatment (S145 and S124) and single CTC sample after treatment (S094), they all carried CREBBP and EGFR mutation, which were frequently showed in many cancers. EWSR1 and TP53 gene mutation both appeared in two samples respectively, while none of the above

10 genes was found mutated in the ctDNA sample after treatment (S63). Additionally, we evaluated the SNVs, InDels and gene mutations shared between these 4 samples, and the Venn diagram was used to illustrate these results. As shown in Figure 6, there were 4 common mutation sites between tissue and ctDNA samples before treatment (S145 and S124), affecting four known tumor-related genes CREBBP, ROS1, TP53 and EGFR, which might help with the selection of treatment options. Moreover, oncogene HRAS mutated both in single CTC sample and ctDNA sample after treatment (S094 and S063) rather than samples before treatment, suggesting the possible relevance between this gene mutation and targeted therapy.

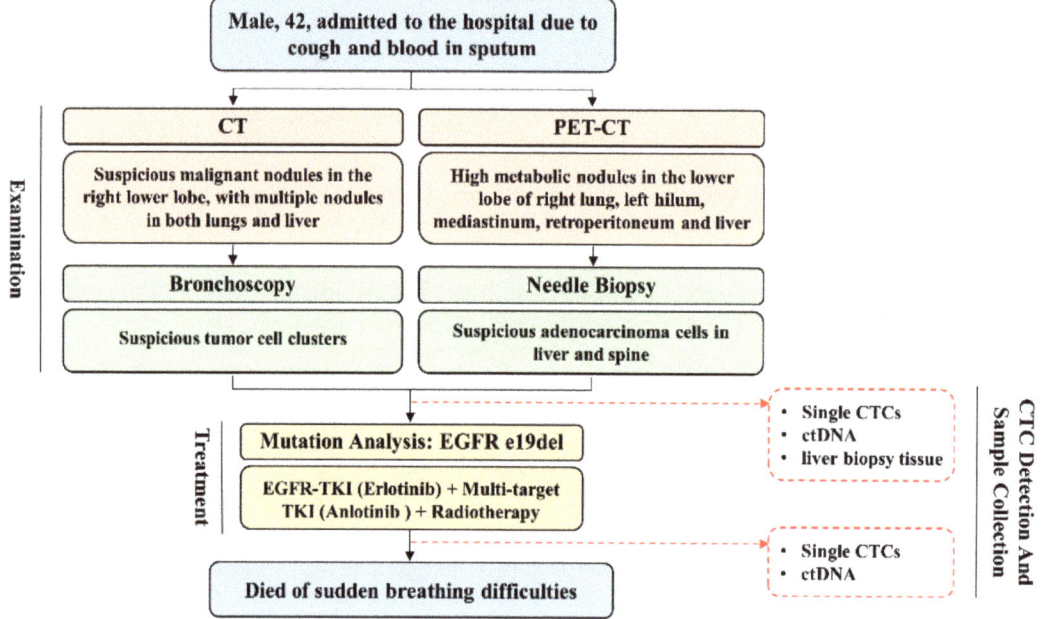

Figure 4. Schematic illustration of the diagnosis, treatment and sample collection of a representative advanced lung cancer patient (patient #1).

Table 2. Sequencing information of samples from patient #1.

Sample ID	Sample Type	Sequencing Technique	Number of SNVs	Number of InDels
S145	Liver biopsy tissue	Whole exome sequencing	132	4
S124	ctDNA before treatment	HapOnco-605 gene panel	8	2
S090	3 single CTCs before treatment	HapOnco-605 gene panel	–	–
S063	ctDNA after treatment	HapOnco-605 gene panel	3	3
S094	1 single CTC after treatment	HapOnco-605 gene panel	1187	177

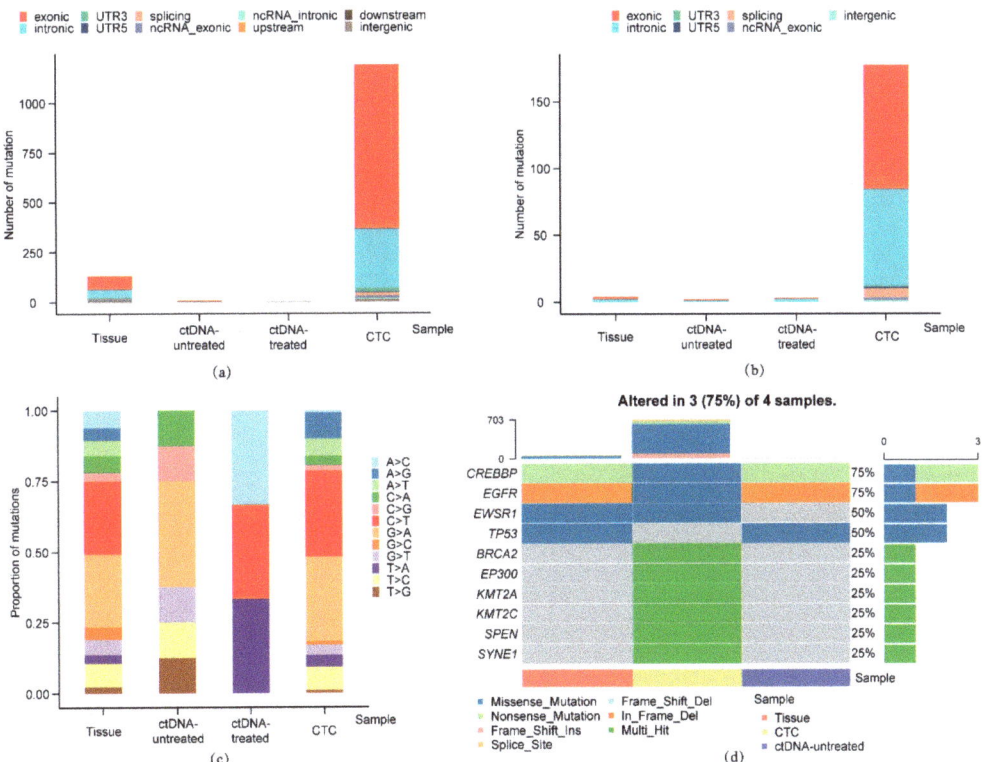

Figure 5. The genetic variations of liver biopsy tissue, ctDNA and single CTC samples. (**a**) The distribution of SNVs on the genome. (**b**) The distribution of InDels on the genome. (**c**) The proportion of different base mutation types. (**d**) The top 10 somatic mutations among 4 samples.

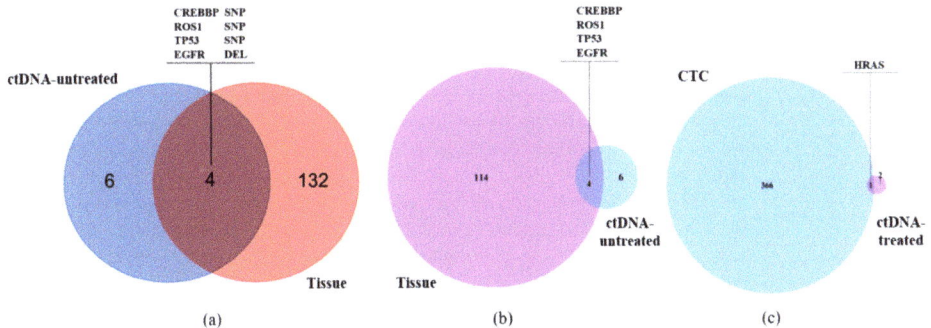

Figure 6. Common SNVs, InDels and mutated genes in liver biopsy tissue, ctDNA and single CTC samples. (**a**,**b**) Common mutation sites and related genes in ctDNA and liver biopsy tissue samples before treatment (S124 and S145). (**c**) Common gene mutation in single CTC and ctDNA sample after treatment (S094 and S063). SNP: single nucleotide polymorphism; DEL: deletion.

4. Discussion

CTCs are considered to be responsible for the distant metastasis of tumors [21]. As a liquid biopsy, CTC enumeration could predict disease progression and treatment response, while molecular analysis may provide deep insights into mechanisms of tumor metastasis and guide the development of therapeutic strategies [22–24]. In recent years, advancements in microfluidic techniques have provided us lots of new tools for cellular biology studies. For example, Pagliara et al. revealed the importance of nuclear structure of embryonic stem cells during their differentiation with microfluidic assay [25], while Cama et al. successfully achieved rapid detection of antibiotic concentration in bacteria by developing a microfluidic platform [26]. Furthermore, the upgradation of materials and refinement of processing have made single-cell studies possible. The combination of microfluidics with imaging equipment have elucidated patterns of long-term growth, division and ageing of Escherichia coli cells [27,28]. While for CTC studies, there have also been many microfluidic devices established to enable more comprehensive understanding of CTCs from a single cell perspective, such as single-CTC metabolic subtypes identification and phenotypes tracking, even targeted proteomics [29–31]. However, single-CTCs analysis, especially when combined with high-throughput sequencing, remains a great challenge due to the lack of technologies to obtain intact CTCs qualified for sequencing with high throughput, high efficiency and less cell damage.

In this work, we reported an integrated microfluidic device for CTCs isolation and further single-cell analysis. We observed a capture rate of around 85%, which basically reached the efficiency when using the membrane alone in our previous study [17]. The capture purity of this platform was 60.4% when cell concentration was 10^4/mL, which was significantly improved than before [17,18]. Then, to test the performance of our device in the clinic, we analyzed blood samples of 16 lung cancer patients and 4 non-tumor patients. We found CTCs in 14 of 16 cancer patients with a positive detectable rate of 87.5%. For one case where we failed to detect CTCs, the medical record revealed that this patient had received surgery and there was no sign of tumor recurrence by the time of detection, which could illustrate the absence of CTCs in his peripheral blood sample. It is worth mentioning that DNA sequencing was performed on single CTCs, ctDNA and liver biopsy tissue of a representative advanced lung cancer patient. Through bioinformatics analysis, we found CTCs hold the most abundant and most complex genetic variations (SNVs and InDels), while ctDNA samples had the least. More importantly, we identified EGFR mutation in liver biopsy tissue and ctDNA sample before treatment, as well as single CTCs after treatment, not in ctDNA sample after treatment. These results suggested that CTCs might be better sources of liquid biopsy than ctDNA, both for tumor heterogeneity research and clinical treatment options. Interestingly, we only found common gene mutations in pre-treatment and post-treatment samples, which indicated the strong impact of treatment on genetic profile. Nevertheless, we cannot draw definitive conclusions relevant to tumor metastasis or treatment response from the current sequencing data, due to the tumor heterogeneity. More studies are needed and the sample size should be expanded to establish the relationship between genetic information and tumor progression. Besides, we are collecting cases similar to patient #1, to get more genetic data from liquid biopsy samples, single CTCs in particular. Findings from this study will help to elucidate the mechanisms of lung cancer liver metastasis.

In this work, we have overcome some limitations addressed in our previous study. First, we replaced the DLD structure with a micropore-arrayed filtration membrane. This change significantly increased CTCs separation throughput of this device without any driving force, such as the pump we used before, while reducing the damage to CTCs caused by fluid shear force and collision with micro-posts. Second, the combination of the filtration system and the CD45 dynabeads based purifying unit dramatically improved the capture purity than using the filtration membrane alone [17], which would decrease the interference of WBCs during single-cell isolation.

However, there remain some difficulties to be addressed. As we can see, the capture rate of this device (around 85%) is slightly lower than using the filtration membrane alone in our previous experiments (83.2–86.7%) [17]. The possible cause of this problem is the cell loss during backwashing, for the cell loss rate of this step in our current work is about 7.3–11.8% (Table S3). To improve this, we are focusing on surface modification of the filtration membranes with biocompatible materials to achieve the controlled recovery of CTCs. Moreover, the operation of this device requires manual manipulation, which greatly impairs sample processing efficiency. For this issue, we have started to upgrade and automate this device, expecting to achieve translation and clinical application.

To summarize, we have established an integrated platform for CTCs detection and isolation with high throughput and low cell damage. Combining with single-cell methods, our work provides a promising tool for cancer research from a new perspective and paves the way for the implementation of CTC analysis into routine clinical practice.

Supplementary Materials: The following are available online at https://www.mdpi.com/2072-666X/12/1/49/s1, Table S1: Capture rate of the integrated device under different concentrations of PC-9 cells. Table S2: Capture purity of the integrated device at a concentration of 10^4 cells/mL of PC-9 cells. Table S3: Cell loss rate of the backwashing step under different concentrations of PC-9 cells.

Author Contributions: Data curation, X.Z.; funding acquisition, Q.W.; resources, E.L.; validation, S.W.; writing—original draft, M.X.; writing—review and editing, W.L. and K.Z.; M.X., W.L. and K.Z. contributed equally. All authors have read and agreed to the published version of the manuscript.

Funding: This research was funded by National Natural Science Foundation of China, grant number 81972916 and 81502702.

Informed Consent Statement: Informed consent was obtained from all subjects involved in the study.

Acknowledgments: We thank all patients and healthy volunteers who participated in this study. We thank Chong Zhang from Oral and Maxillofacial Disease Institute, The Second Hospital Affiliated to Dalian Medical University for assistance with schematic diagram and image processing.

Conflicts of Interest: All authors declare no conflict of interest.

References

1. Paget, S. The distribution of secondary growths in cancer of the breast. *Cancer Metastasis Rev.* **1989**, *8*, 98–101. [PubMed]
2. Kilgour, E.; Rothwell, D.G.; Brady, G.; Dive, C. Liquid Biopsy-Based Biomarkers of Treatment Response and Resistance. *Cancer Cell* **2020**, *37*, 485–495. [CrossRef] [PubMed]
3. Malone, E.R.; Oliva, M.; Sabatini, P.J.B.; Stockley, T.L.; Siu, L.L. Molecular profiling for precision cancer therapies. *Genome Med.* **2020**, *12*, 8. [CrossRef] [PubMed]
4. Gossett, D.R.; Tse, H.T.; Lee, S.A.; Ying, Y.; Lindgren, A.G.; Yang, O.O.; Rao, J.; Clark, A.T.; Di Carlo, D. Hydrodynamic stretching of single cells for large population mechanical phenotyping. *Proc. Natl. Acad. Sci. USA* **2012**, *109*, 7630–7635. [CrossRef]
5. Hodgson, A.C.; Verstreken, C.M.; Fisher, C.L.; Keyser, U.F.; Pagliara, S.; Chalut, K.J. A microfluidic device for characterizing nuclear deformations. *Lab Chip* **2017**, *17*, 805–813. [CrossRef]
6. Bamford, R.A.; Smith, A.; Metz, J.; Glover, G.; Titball, R.W.; Pagliara, S. Investigating the physiology of viable but non-culturable bacteria by microfluidics and time-lapse microscopy. *BMC Biol.* **2017**, *15*, 121. [CrossRef]
7. Vergalli, J.; Dumont, E.; Pajović, J.; Cinquin, B.; Maigre, L.; Masi, M.; Réfrégiers, M.; Pagès, J.M. Spectrofluorimetric quantification of antibiotic drug concentration in bacterial cells for the characterization of translocation across bacterial membranes. *Nat. Protoc.* **2018**, *13*, 1348–1361. [CrossRef]
8. Liu, A.L.; He, F.Y.; Wang, K.; Zhou, T.; Lu, Y.; Xia, X.H. Rapid method for design and fabrication of passive micromixers in microfluidic devices using a direct-printing process. *Lab Chip* **2005**, *5*, 974–978. [CrossRef]
9. Cama, J.; Chimerel, C.; Pagliara, S.; Javer, A.; Keyser, U.F. A label-free microfluidic assay to quantitatively study antibiotic diffusion through lipid membranes. *Lab Chip* **2014**, *14*, 2303–2308. [CrossRef]
10. Mathur, L.; Ballinger, M.; Utharala, R.; Merten, C.A. Microfluidics as an Enabling Technology for Personalized Cancer Therapy. *Small* **2020**, *16*, e1904321. [CrossRef]
11. Lawson, D.A.; Kessenbrock, K.; Davis, R.T.; Pervolarakis, N.; Werb, Z. Tumour heterogeneity and metastasis at single-cell resolution. *Nat. Cell Biol.* **2018**, *20*, 1349–1360. [CrossRef] [PubMed]
12. Keller, L.; Pantel, K. Unravelling tumour heterogeneity by single-cell profiling of circulating tumour cells. *Nat. Rev. Cancer* **2019**, *19*, 553–567. [CrossRef] [PubMed]

13. Morita, K.; Wang, F.; Jahn, K.; Hu, T.; Tanaka, T.; Sasaki, Y.; Kuipers, J.; Loghavi, S.; Wang, S.A.; Yan, Y.; et al. Clonal evolution of acute myeloid leukemia revealed by high-throughput single-cell genomics. *Nat. Commun.* **2020**, *11*, 5327. [CrossRef] [PubMed]
14. McFarland, J.M.; Paolella, B.R.; Warren, A.; Geiger-Schuller, K.; Shibue, T.; Rothberg, M.; Kuksenko, O.; Colgan, W.N.; Jones, A.; Chambers, E.; et al. Multiplexed single-cell transcriptional response profiling to define cancer vulnerabilities and therapeutic mechanism of action. *Nat. Commun.* **2020**, *11*, 4296. [CrossRef] [PubMed]
15. Mansilla, C.; Soria, E.; Ramírez, N. The identification and isolation of CTCs: A biological Rubik's cube. *Crit. Rev. Oncol. Hematol.* **2018**, *126*, 129–134. [CrossRef] [PubMed]
16. Yu, L.; Sa, S.; Wang, L.; Dulmage, K.; Bhagwat, N.; Yee, S.S.; Sen, M.; Pletcher, C.H., Jr.; Moore, J.S.; Saksena, S.; et al. An integrated enrichment system to facilitate isolation and molecular characterization of single cancer cells from whole blood. *Cytom. A* **2018**, *93*, 1226–1233. [CrossRef]
17. Liu, Y.; Li, T.; Xu, M.; Zhang, W.; Xiong, Y.; Nie, L.; Wang, Q.; Li, H.; Wang, W. A high-throughput liquid biopsy for rapid rare cell separation from large-volume samples. *Lab Chip* **2018**, *19*, 68–78. [CrossRef]
18. Jiang, J.; Zhao, H.; Shu, W.; Tian, J.; Huang, Y.; Song, Y.; Wang, R.; Li, E.; Slamon, D.; Hou, D.; et al. An integrated microfluidic device for rapid and high-sensitivity analysis of circulating tumor cells. *Sci. Rep.* **2017**, *7*, 42612. [CrossRef]
19. Xu, M.; Zhao, H.; Chen, J.; Liu, W.; Li, E.; Wang, Q.; Zhang, L. An Integrated Microfluidic Chip and Its Clinical Application for Circulating Tumor Cell Isolation and Single-Cell Analysis. *Cytom. A* **2020**, *97*, 46–53. [CrossRef]
20. Hao, S.J.; Wan, Y.; Xia, Y.Q.; Zou, X.; Zheng, S.Y. Size-based separation methods of circulating tumor cells. *Adv. Drug Deliv. Rev.* **2018**, *125*, 3–20. [CrossRef]
21. Dianat-Moghadam, H.; Azizi, M.; Eslami-S, Z.; Cortés-Hernández, L.E.; Heidarifard, M.; Nouri, M.; Alix-Panabières, C. The Role of Circulating Tumor Cells in the Metastatic Cascade: Biology, Technical Challenges, and Clinical Relevance. *Cancers* **2020**, *12*, 867. [CrossRef] [PubMed]
22. Franses, J.W.; Philipp, J.; Missios, P.; Bhan, I.; Liu, A.; Yashaswini, C.; Tai, E.; Zhu, H.; Ligorio, M.; Nicholson, B.; et al. Pancreatic circulating tumor cell profiling identifies LIN28B as a metastasis driver and drug target. *Nat. Commun.* **2020**, *11*, 3303. [CrossRef] [PubMed]
23. Klotz, R.; Thomas, A.; Teng, T.; Han, S.M.; Iriondo, O.; Li, L.; Restrepo-Vassalli, S.; Wang, A.; Izadian, N.; MacKay, M.; et al. Circulating Tumor Cells Exhibit Metastatic Tropism and Reveal Brain Metastasis Drivers. *Cancer Discov.* **2020**, *10*, 86–103. [CrossRef] [PubMed]
24. Zhong, X.; Zhang, H.; Zhu, Y.; Liang, Y.; Yuan, Z.; Li, J.; Li, J.; Li, X.; Jia, Y.; He, T.; et al. Circulating tumor cells in cancer patients: Developments and clinical applications for immunotherapy. *Mol. Cancer* **2020**, *19*, 15. [CrossRef]
25. Pagliara, S.; Franze, K.; McClain, C.R.; Wylde, G.; Fisher, C.L.; Franklin, R.J.M.; Kabla, A.J.; Keyser, U.F.; Chalut, K.J. Auxetic nuclei in embryonic stem cells exiting pluripotency. *Nat. Mater.* **2014**, *13*, 638–644. [CrossRef] [PubMed]
26. Cama, J.; Voliotis, M.; Metz, J.; Smith, A.; Iannucci, J.; Keyser, U.F.; Tsaneva-Atanasova, K.; Pagliara, S. Single-cell microfluidics facilitates the rapid quantification of antibiotic accumulation in Gram-negative bacteria. *Lab Chip* **2020**, *20*, 2765–2775. [CrossRef]
27. Wang, P.; Robert, L.; Pelletier, J.; Dang, W.L.; Taddei, F.; Wright, A.; Jun, S. Robust growth of Escherichia coli. *Curr. Biol.* **2010**, *20*, 1099–1103. [CrossRef]
28. Łapińska, U.; Glover, G.; Capilla-Lasheras, P.; Young, A.J.; Pagliara, S. Bacterial ageing in the absence of external stressors. *Philos. Trans. R. Soc. Lond. B Biol. Sci.* **2019**, *374*, 20180442. [CrossRef]
29. Brisotto, G.; Biscontin, E.; Rossi, E.; Bulfoni, M.; Piruska, A.; Spazzapan, S.; Poggiana, C.; Vidotto, R.; Steffan, A.; Colombatti, A.; et al. Dysmetabolic Circulating Tumor Cells Are Prognostic in Metastatic Breast Cancer. *Cancers (Basel).* **2020**, *12*, 1005. [CrossRef]
30. Poudineh, M.; Aldridge, P.M.; Ahmed, S.; Green, B.J.; Kermanshah, L.; Nguyen, V.; Tu, C.; Mohamadi, R.M.; Nam, R.K.; Hansen, A.; et al. Tracking the dynamics of circulating tumour cell phenotypes using nanoparticle-mediated magnetic ranking. *Nat. Nanotechnol.* **2017**, *12*, 274–281. [CrossRef]
31. Sinkala, E.; Sollier-Christen, E.; Renier, C.; Rosàs-Canyelles, E.; Che, J.; Heirich, K.; Duncombe, T.A.; Vlassakis, J.; Yamauchi, K.A.; Huang, H.; et al. Profiling protein expression in circulating tumour cells using microfluidic western blotting. *Nat. Commun.* **2017**, *8*, 14622. [CrossRef] [PubMed]

Article

Single Cell Hydrodynamic Stretching and Microsieve Filtration Reveal Genetic, Phenotypic and Treatment-Related Links to Cellular Deformability

Fenfang Li [1,2,*], Igor Cima [3,4], Jess Honganh Vo [3,5], Min-Han Tan [3,5] and Claus Dieter Ohl [2,6,*]

1. Lee Kong Chian School of Medicine, Nanyang Technological University Singapore, 11 Mandalay Road, Singapore 308232, Singapore
2. Division of Physics and Applied Physics, School of Physical and Mathematical Sciences, Nanyang Technological University Singapore, 21 Nanyang Link, Singapore 637371, Singapore
3. Institute of Bioengineering and Nanotechnology, 31 Biopolis Way, Singapore 138669, Singapore; i.cima@dkfz.de (I.C.); jess.vo@lucence.com (J.H.V.); minhan.tan@lucence.com (M.-H.T.)
4. DKFZ-Division Translational Neurooncology at the WTZ, DKTK partner site, University Hospital Essen, 45147 Essen, Germany
5. Lucence Diagnostics Pte Ltd., 211 Henderson Road, Henderson Industrial Park, Singapore 159552, Singapore
6. Institute for Physics, Faculty of Natural Sciences, Otto-von-Guericke University of Magdeburg, 39106 Magdeburg, Germany
* Correspondence: fli2@e.ntu.edu.sg (F.L.); claus-dieter.ohl@ovgu.de (C.D.O.)

Received: 30 March 2020; Accepted: 8 May 2020; Published: 9 May 2020

Abstract: Deformability is shown to correlate with the invasiveness and metastasis of cancer cells. Recent studies suggest epithelial-to-mesenchymal transition (EMT) might enable cancer metastasis. However, the correlation of EMT with cancer cell deformability has not been well elucidated. Cellular deformability could also help evaluate the drug response of cancer cells. Here, we combine hydrodynamic stretching and microsieve filtration to study cellular deformability in several cellular models. Hydrodynamic stretching uses extensional flow to rapidly quantify cellular deformability and size with high throughput at the single cell level. Microsieve filtration can rapidly estimate relative deformability in cellular populations. We show that colorectal cancer cell line RKO with the mesenchymal-like feature is more flexible than the epithelial-like HCT116. In another model, the breast epithelial cells MCF10A with deletion of the TP53 gene are also significantly more deformable compared to their isogenic wildtype counterpart, indicating a potential genetic link to cellular deformability. We also find that the drug docetaxel leads to an increase in the size of A549 lung cancer cells. The ability to associate mechanical properties of cancer cells with their phenotypes and genetics using single cell hydrodynamic stretching or the microsieve may help to deepen our understanding of the basic properties of cancer progression.

Keywords: cancer metastasis; deformability; epithelial to mesenchymal transition; TP53 genes; chemotherapy drug; microfluidic hydrodynamic stretching; microsieve

1. Introduction

Invasion of cancer cells into the blood stream is an essential step for their metastatic spread from primary tumor to distant organs [1]. Recent studies suggest that the epithelial-to-mesenchymal transition (EMT) might enable metastatic dissemination by generating more migratory and invasive cells [2–4]. EMT is a fundamental program for embryonic development and differentiation of tissues/organs, where conversions of epithelial cells to mesenchymal cells occur. However, EMT can also promote tumor progression by generating cancer stem cells that allow phenotypic changes, where cells decrease cell-cell adhesion and increase their motility and invasiveness [2–4].

Biomechanical properties such as deformability have been shown to correlate with the invasion potential of cancer cells [5,6]. However, the correlation of EMT with deformability of cancer cells has not been well studied and remains elusive. Higher deformability has been reported for pancreatic cancer cells undergoing EMT [7]. Some circulating tumor cells (CTCs) from metastatic prostate cancer patients also exhibit smaller size and increased flexibility similar to blood cells rather than typical tumor cell lines [8]. However, the latter study also shows evidence that for cancer cells to be able to exit a tumor and enter circulation they are not required to be more deformable than the cells that were first injected into the tumor. Moreover, for TP53 genetic alterations that are demonstrated to cause EMT [9], their effect on the deformability of cancer cells has not been studied.

Additionally, there is a growing interest to understand the relationship between pharmacology and cancer cell stiffness [10,11]. Increasing evidence shows invasive cancer cells have higher cell deformability compared to benign or normal cells of the same origin. Thus, cell deformability could serve as a promising label-free biomarker for the underlying cytoskeletal or nuclear changes that are associated with disease processes and change in cell state, especially under the intervention of anticancer drugs. For example, for cellular sensitive response, perceptible differences in cell deformability before and after drug treatment are expected, whereas cells with resistive response are expected to have little changes in cellular deformability pre- and posttreatment. Therefore, a method capable of measuring the deformability for a large population of cells from cancer biopsies pre- and post-chemotherapeutic drug treatment would aid clinical drug screening for personalized medicine.

Conventional approaches to measure single-cell mechanical properties include magnetic twisting cytometry [12], optical stretcher [13], atomic force microscope [14], micropipette aspiration [15] and cell transit analyzer [16]. However, these methods are time consuming for testing a large number of cells to address the inherent biological variation and heterogeneity within a cell population. A recently developed method called controlled cavitation rheology (CCR) allows simultaneous measurement of the deformability for a large number of red blood cells in hundreds of microseconds with single cell resolution [17–19]. However, little cellular deformation can be induced with this method for spherical cancer cells in suspension [20].

More recently, high-throughput techniques for cell deformability measurement have been developed, such as real-time deformability cytometry [21], inertial microfluidic cell stretcher (iMCS) [22] and hydrodynamic stretching [23]. The hydrodynamic stretching method utilizes a simple setup with a microscope, syringe pump and high speed camera. It enables measuring both the deformability and size for hundreds of cells per second, where single cells in suspension are delivered by inertial focusing to the center of a microfluidic cross channel and get stretched by an extensional flow [23]. The high throughput and automation of analysis allow us to rapidly link cellular deformability with various phenotypes at higher statistical significance than conventional methods. In addition, we also compare the results from this method with those from the recently developed microsieve device. This silicon microsieve has been developed for label-free and rapid isolation of circulating tumor cells (CTCs) from whole blood samples by utilizing membrane-based filtration [24]. It has the advantage of excellent definition of pore size together with a high pore density that allows for high-throughput sample processing. Recently, the microsieve has also been applied for the rapid screening of chemotherapy drugs by comparing the retrieve ratio of untreated vs. treated carcinoma cell lines. Still, the underlying effects of the chemotherapy drugs are unclear as both the cell size and deformability may affect cells passing through the microsieve. Moreover, the heterogeneity within the cell population cannot be resolved by the microsieve method due to a lack of single cell resolution. Thus, the hydrodynamic stretching method has the potential to unravel the separation mechanism for a specific cell sample of the microsieve method, e.g., difference in size or deformability, which can in turn help to choose the right pore size for the isolation.

To achieve these goals, we combine the methods of microfluidic hydrodynamic stretching and microsieve to characterize the differential deformability of several cell models related to increased invasiveness, EMT or chemotherapy drug treatment. In the study we use cell models that were reported to exhibit more invasive mesenchymal features (human colon carcinoma cell line RKO) [25,26] and undergo EMT (MCF10A TP53 knockout) [9] compared to their counterparts (HCT116 and MCF10A wildtype (wt), respectively), as well as the carcinoma cell line A549 treated with cytochalasin D or the chemotherapy drug docetaxel, which are believed to change cellular stiffness through alteration of cell cytoskeleton.

2. Materials and Methods

2.1. Microfluidic Device and Hydrodynamic Stretching Method

A schematic of the individual microchannel is shown in Figure 1A. An array of micro-filters is designed after the inlet to block cell aggregates or other debris to avoid clogging the channel. Asymmetric curvatures result in inertial focusing of the cells towards the channel center. The solution containing the cells flows from top and bottom towards the stretching region located in the channel center, as seen in the schematic in Figure 1B. Cells approaching the stagnation point at the center will experience an elongational flow and are stretched perpendicular to the direction of the streamline. As the flow is symmetric, cells may reach the stretching region from below and above (see Figure 1B). Cells leave the stretching region through either one of the two outlets (left and right) where they quickly recover their shape. The channel height is 26 µm and the channel width before and after the stretching region is 67 µm. Eleven of these microchannels fit on a single glass slide with size 75 × 50 mm².

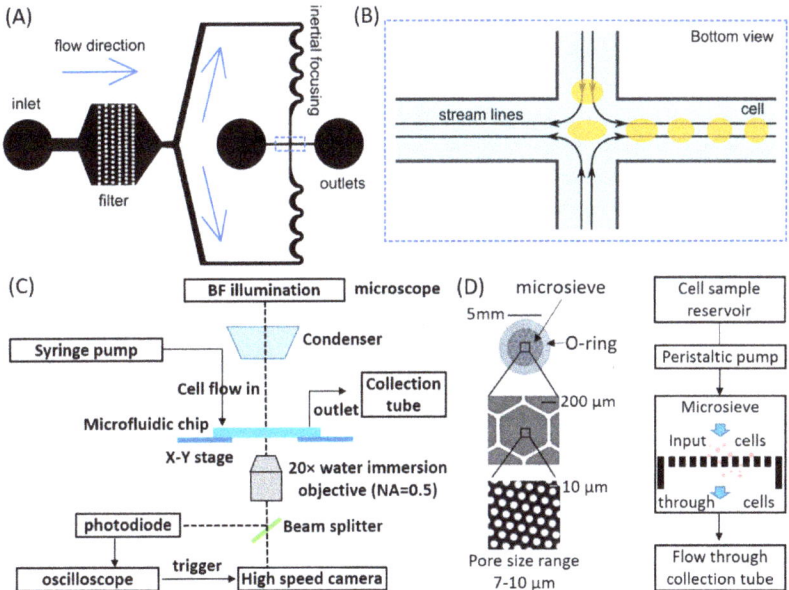

Figure 1. The experimental devices and the working principle. (**A**) A schematic of the individual hydrodynamic stretching microchannel (channel height = 26 µm). The channel consists of one inlet, two outlets and a filtering region to avoid cell aggregates and debris. The flow enters from above and below the central stretching region (marked by the dashed box). Cells are focused to the channel center line by the inertial focusing curvatures and are deformed in the stretching region. (**B**) A sketch of the stretching region, which is marked by the dashed box in A. (**C**) A schematic diagram showing the operation of the hydrodynamic stretching microchannel and image recording. (**D**) The structure of a microsieve and its working principle.

We used standard soft lithography techniques to fabricate the patterned SU-8 based master mold, from which the polydimethylsiloxane (PDMS, Sylgard 184 Silicone Elastomer Kit, Dow Corning, Midland, MI, USA) microchannel was cast. The PDMS microchannel was then bonded onto the glass slide (75 × 50 mm^2) immediately after 45 s treatment of oxygen plasma using a plasma cleaner (Expanded Plasma Cleaner PDC-002, Harrick Plasma, Ithaca, NY, USA). When the input flow rate was around several hundred μL/min, we glued the inlet with a fast setting epoxy adhesive (Araldite) to avoid liquid leakage. An input cell concentration of 1.32×10^5–2.64×10^5 cells/mL was used for the hydrodynamic stretching experiment.

2.2. Experimental Operation for Hydrodynamic Stretching

The microfluidic device is placed on an inverted microscope (IX-71, Olympus, Tokyo, Japan) and imaged with a water immersion objective (20×/0.50 NA, Olympus). A syringe loaded with the cell solution is placed in an infusion pump (KDS200, KD Scientific, Holliston, MA, USA) operating at an optimized flow rate of 500–700 μL/min. The syringe is connected to the channel inlet through a microbore tubing (Tygon ND-100-80) that has an inner diameter of 0.508 mm and an outer diameter of 1.524 mm. The two channel outlets are plugged with 90° bent blunt needles (SH23-B-90, SAN-EI Tech Asia Pte Ltd., Singapore) and are connected to a collection reservoir with tubing of equal lengths. A photodiode (DET 10A/M, Thorlabs, Newton, NJ, USA) is placed at the image plane of the microscope side port and is connected to an oscilloscope (Lecroy, wave runner 64 Xi-A). The sensor of the photodiode is aligned with the image of the channel center (stretching region) shown in Figure 1B. When a cell passes through the stretching region, there is a fluctuation of the transmitted light intensity at the cell location. This can be detected by the photodiode and sent to the oscilloscope, which uses the signal to trigger image capturing using a high-speed camera (Photron SA-X). The high-speed camera is connected to the trinocular through a 0.5× demagnifying C-mount. The camera is set to the center trigger mode so that it will capture images both before and after the electronic trigger sent. In this way, the high-speed camera records the passing of cells only and thus greatly reduces the file size of the high-speed recordings. The typical operation procedure is as follows: after the syringe has infused the cell solution, we wait about 15–20 s for the flow to reach a steady state and then start the high-speed image recording at a frame rate of 180,000 fps and an exposure time of 293 ns. The size of the high-speed camera view field is 512 × 208 μm. A schematic diagram for the experimental setup is shown in Figure 1C.

2.3. Microsieve Method

In this method, the cell suspension is pumped with a peristaltic pump through a densely packed array of pores, i.e., the silicon microsieve. For a specific pore size (7–10 μm), smaller or more flexible cells can pass through the microsieve while larger or more rigid cells will be rejected (Figure 1D). We tested various flow rates and found that 0.5–1 mL/min is the optimal flow rate to achieve deformation of the cells and avoid cell damage when going through the pores. An input cell concentration of 1×10^3–2.5×10^3 cells/mL is used for the microsieve filtration experiment. For more details about the fabrication of the microsieve and its operation, please refer to [24] and [27], respectively.

2.4. Image Processing and Data Analysis

We have developed a Matlab script to process the recorded images from hydrodynamic stretching. The details of the image processing and data analysis are illustrated with an example of a single RKO cell in Figure 2. Figure 2A reveals that the cell is progressively elongated when it enters the extensional stretching region (0–105.6 μs). Once it flows toward the left outlet its elongated shape recovers gradually (161.1–255.6 μs). To trace the cell, each image is background subtracted from an image frame without cells. The cell's contour is determined through thresholding, after which the cell centroid can be estimated. The cell trajectory is the time-dependent curve of its centroid (Figure 2B). After the cell centroid is found, we can define a circular bounding box (blue) around the cell (Figure 2C).

Then, we use an energy minimization algorithm called active contour to obtain the precise contour of the cell shape (red), and further get its center of mass (a more precise centroid) and major (a) and minor (b) semi-axis. The active contour method detects the edge of the cell using the intensity gradient (the cell edge is darker compared to the outside and inside of the cell, thus has the highest intensity gradient) [28]. It can track the cell edge more accurately and smoothly than the thresholding method as the results of the latter can be easily affected by the fluctuation of the pixel intensity due to camera noise and illumination.

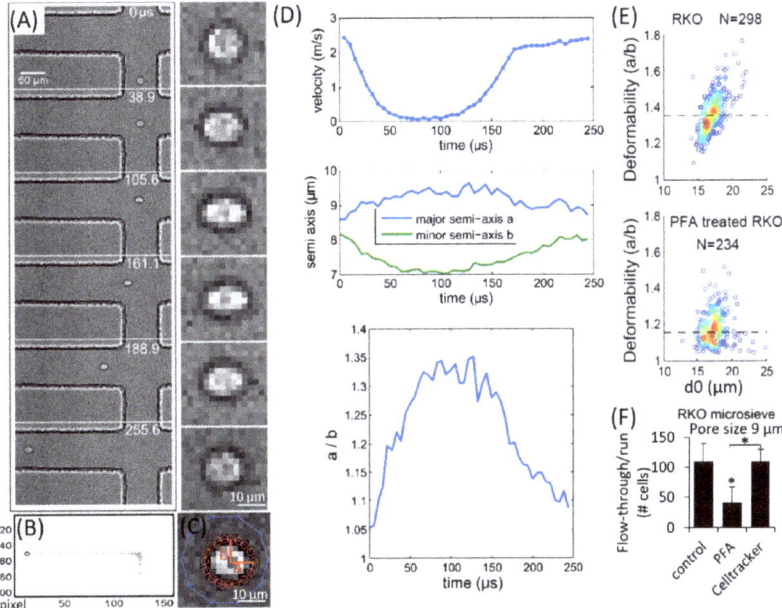

Figure 2. Illustration of image processing and data analysis with an exemplary RKO cell. (**A**) Left column: selected image sequences for a single cell traveling through the stretching extensional flow region. Right column: enlarged images of the cell shown in the left column. (**B**) Cell detection (*circle*) with background subtraction and thresholding, showing trajectory of cell centroid (*cross symbols*) across the channel. (**C**) Active contour method for extraction of cell contour, major, *a*, and minor, *b*, semi-axis. The red circles denote the cell boundary elements detected using the active contour method. (**D**) Extracted data for the temporal evolution of the cell velocity (*top*), its major (*blue*) and minor (*green*) semi-axis (*middle*), and the ratio between the major and minor semi-axis (*a/b*) (*bottom*). (**E**) Statistical analysis of the cell deformability vs. relaxation diameter d_0 with the density scatter plot for untreated RKO and paraformaldehyde (PFA)-treated RKO cells. The dashed lines indicate the median deformability. A hotter color indicates a higher data density. The deformability is defined as the maximum value of a/b, while d_0 is the averaged diameter when the ratio a/b is minimum. The PFA-treated RKO cells have a significantly lower deformability compared to untreated RKO cells, $p < 0.0001$ from two-tailed student t test. (**F**) Averaged number of cells flowing through the microsieve (pore size 9 µm) per run for non-treated RKO (control), PFA-treated RKO and RKO loaded with cell tracker fluorescence dye with the same input number of cells. Three replicates were done for each microsieve experiment (n = 3). There are significantly fewer flow-through cells for PFA-treated RKO compared to the control group, * $p < 0.05$ from one-way ANOVA test followed by post-hoc Tukey Honest Significant Difference (HSD) test. No significant difference is observed between control and cell tracker loaded group ($p = 0.90$). The error bars are standard deviations from three repeated microsieve measurements.

With the above obtained cell centroids, the averaged interframe cell velocity can be calculated. This is shown in Figure 2D (top frame). During the approach towards the stretching region the velocity decreases from about 2.5 m/s to a minimum of close to 0. Then, the cell leaves the stretching region, while its velocity increases gradually back to a nearly constant value. The corresponding temporal evolution of the cell's semi-axis is shown in the center graph of Figure 2D. From $t = 0$ to $t = 100$ µs the length of the major semi-axis progressively increases while the minor semi-axis decreases, i.e., the cell is elongated. Upon leaving the stretching region this trend is reversed, and the original shape is recovered. The length ratio of the major to minor semi-axis, a/b, is shown at the bottom of Figure 2D. It reaches a maximum value of around $t = 100$ µs when the cell is at the center of the channel crossing. Therefore, we use max(a/b) to define the cell deformability. A set of control experiments of RKO and PFA-treated RKO with hydrodynamic stretching and microsieve separation were performed to validate the workings of the two microfluidic approaches (Figure 2E,F). We followed the previous work [23] and present the cell deformability and cell size as a density scatter plot (see Figure 2E), where d_0 is the averaged diameter calculated from the cell area of the most spherical cell, i.e., when the ratio a/b is smallest.

To make sure we are measuring single cells, we pipetted up and down the cell solution carefully to reduce clumpy cells during sample preparation. When putting them into the chip, we may still have some clumpy cells. Larger clusters can be blocked by the filter array near the chip inlet (Figure 1A). Smaller clusters such as two cells that stuck together can be rejected during real-time visualization of our imaging processing. We checked each cell during the automated imaging processing to ensure it is single cell measurement.

2.5. Cell Culture and Preparation

All cell lines used in this study except MCF10A were cultured in a humidified incubator at 37 °C and 5% CO_2 with culture medium (Dulbecco's modified Eagle's medium (DMEM) with 2.5 mM L-glutamine and 10% (v/v) fetal bovine serum and 1% (v/v) penicillin/ streptomycin).

The culture conditions of MCF10A wildtype and TP53 knockout followed the manufacturer's instructions: the culture medium is made of DMEM/Ham's Nutrient Mixture F12 (1:1) with 2.5 mM L-glutamine, 5% horse serum, 10 mg/mL human insulin, 0.5 mg/mL hydrocortisone, 10 ng/mL EGF and 100 ng/mL cholera toxin. Cells were maintained in a humidified incubator at 37 °C in the presence of 5% CO_2.

2.5.1. HCT116, RKO and PFA-Treated RKO

Two types of human colon carcinoma cell line, HCT116 and RKO, were kept routinely in culture. At around 90% confluence they were split with 0.25% trypsin/EDTA, then diluted with fresh culture medium at a ratio of 1:10 to 1:20 (e.g., 500 µL to 5mL or 10 mL). The cell suspension was gently transferred to a 5mL plastic syringe (BD Bioscience) immediately before the experiment. For the experiments of mixed HCT116 and RKO flowing through a microsieve, tracker red and tracker green (Invitrogen) were used to label HCT116 and RKO, respectively. The total input and passing through cell mixture were characterized using flow cytometry (FACS Calibur instrument from BD).

We used 4% PFA in 1x PBS (sterile filtered) to stiffen the RKO cells. PFA is a common fixative that is used to preserve cell structure. Basically, it creates covalent chemical bonds between proteins, and this anchors soluble proteins to the cytoskeleton, thus, making the cells more rigid. The PFA reagent was added to the RKO cell solution (in fresh culture medium) at a final concentration of 2%. Then, the cells were incubated overnight at 4 °C before they were centrifuged and re-suspended in culture medium for experiments.

2.5.2. MCF10A Wildtype and TP53 Knockout

MCF10A cells are immortalized mammary epithelial cells. MCF10A TP53 knockout and isogenic wildtype cells were purchased as a kit from Sigma (Catalog number CLLS1049) and cultured following the manufacturer's instructions.

2.5.3. A549 Treated with Cytochalasin D and Docetaxel

A549 cells (Adenocarcinomic Human Alveolar Basal Epithelial Cells) treated with cytochalasin D (mycotoxins) (10 μM, 100 μM), docetaxel (a chemotherapy drug) (5, 10, 50 nM) and untreated control cells were prepared for microfluidic hydrodynamic stretching and microsieve tests. For the reagents, cytochalasin D inhibits polymerization and induces depolymerization of actin filaments, while docetaxel inhibits microtubule depolymerization and is assumed to induce cellular stiffness.

2.6. Statistical Analysis

For both hydrodynamic stretching and microsieve results, if only two groups were compared, a two-tailed student t test was used for the statistical analysis of the deformability and cellular size, and the plotted error bars are standard deviation. If more than two groups were compared, a one-way ANOVA test was performed first to check whether one or more treatment groups are significantly different, which is followed by a post-hoc Tukey HSD multiple comparison test to identify which of the pairs of treatments are significantly different from each other.

3. Results

3.1. Control Experiments for Validation of Measurements

To validate the operation and measurements of the microfluidic devices, we performed a control experiment comparing the deformability of non-treated RKO cells with that of PFA-treated RKO cells. The results are shown in Figure 2E,F. Differential deformability between RKO and PFA-treated RKO is observed from the microfluidic hydrodynamic stretching. The PFA treated RKO cells have a significant lower deformability (with averaged a/b ~1.16 compared to 1.36 for untreated RKO cells), which is consistent with the fixation and stiffening effects of PFA. The result is also consistent with the microsieve measurements whereby fewer PFA-treated cells could pass through the microsieve than the non-treated (control) ones. These results suggest that our measurements of the cell deformability are valid.

3.2. Differential Deformability between Cell Phenotypes and Correlation with Epithelial-To-Mesenchymal Transition

3.2.1. HCT116 and RKO

HCT116 and RKO are two different types of human colorectal cancer cell (CRC) lines. Their sizes are different. RKO have a slightly larger averaged diameter of ~17.2 μm as compared to HCT116 with approx. 15.3 μm (Figure 3A). However, when flowing these cell suspensions through the microsieve system, a lower retrieval (higher passing through) is observed for the larger RKO cells as compared to HCT116 (Figure 3B). Particularly, when the pore size of the microsieve increases to around 9 μm (below the averaged diameter for both cell lines), most of the RKO cells can pass through at a high flow rate of 1 mL/min, while the majority of HCT116 cells are trapped on the microsieve. We further mixed both cell lines together and passed them through a microsieve. Prior to that, the RKO cells were labelled with green fluorescent cell tracker while the HCT116 cells were labelled with red fluorescent cell tracker. The concentration of cell tracker used was 0.5 to 1 μM. We also did control experiments and found at these concentrations the cell tracker did not modify the cellular deformability (see Figure 2F). Results from flow cytometry reveal that the mixture consists of 60% RKO and 40% HCT116. Yet 92% of the cells that have passed through the microsieve are RKO cells and the remaining 8% are HCT116 cells (Figure 3C). Interestingly, RKO can pass through the microsieve pores more easily than HCT116.

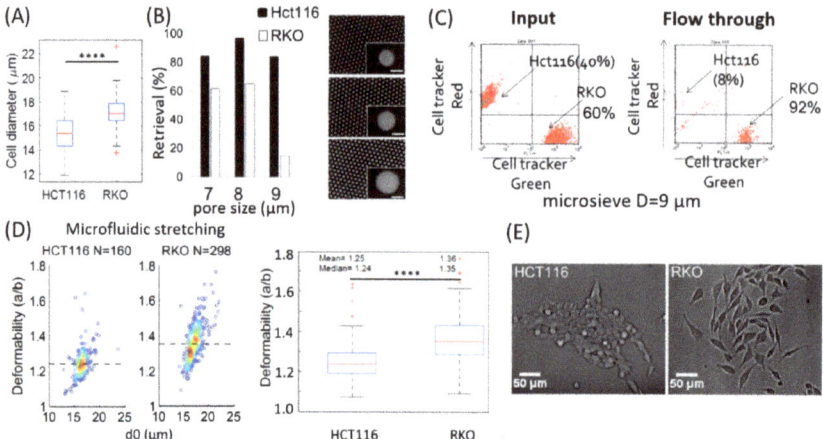

Figure 3. Differential deformability and characteristics of colorectal cancer cell lines HCT116 and RKO. (**A**) Boxplot and statistic results for the diameter of HCT116 and RKO cell lines. Cell line diameters were extracted from images of freshly trypsinized cells in suspension using the software ImageJ. The cell images are taken from an inverted microspore (Olympus IX81) in bright field modus. The line inside the boxes represents the median of the data, while the edges of the boxes are the 25th and 75th percentiles, and the ends of the whiskers denote 1.5 times the interquartile deviation (IQR). The plus symbols on the top or bottom of the whiskers indicate data not included between the whiskers. **** $p < 0.0001$ from two-tailed student t test. (**B**) Retrieval (normalized as the percentage of cells left over on the microsieve surface in total input cells) of HCT116 and RKO cells from microsieves with a pore diameter of 7, 8, and 9 µm (shown on the right side, scale bars denote 7 µm), respectively. (**C**) Percentage of both cell lines in input and after flowing through the microsieve at high flow rate of 1 mL/min and identified with fluorescent cell tracker. (**D**) HCT116 and RKO from microfluidic hydrodynamic stretching: (left) density scatter plot of the deformability. The dashed lines indicate the median deformability. (right) Statistic analysis (the boxes). The HCT116 cells have a significantly lower deformability compared to RKO cells, **** $p < 0.0001$ from two-tailed student t test. (**E**) Bright field images of the cell morphology during adhesion and growth of HCT116 and RKO under the same culture condition.

Next, both cell lines were tested for their deformability with the microfluidic hydrodynamic stretching. The results in Figure 3D reveal a significantly higher deformability of RKO cells, with a median deformability of 1.35 compared to 1.24 for HCT116 cells. That may explain why the RKO cells can pass through the microsieve more easily although they have a larger size as compared to HCT 116.

When making a closer inspection of the cell morphology, as shown in Figure 3E, we can see clearly that HCT116 cells tend to stick to each other during growth, which is a typical epithelial characteristic. In contrast, the RKO cells are loosely arranged as individual cells, and they spread to a larger extent on the petri dish. It indicates that RKO presents similarities to mesenchymal cell phenotype. This is consistent with previous studies showing RKO cells primarily exhibit mesenchymal features and have high invasion potential [25,26], while HCT116 is reported to demonstrate epithelia features [29]. The above results suggest that the CRC mesenchymal cell phenotype correlates with higher deformability compared to the CRC epithelial phenotype from the same origin.

3.2.2. MCF10A Wildtype and Knockout of TP53 Gene

Differential deformability is also observed upon the knockout of the tumor suppressor TP53 gene from the MCF10A human breast cell line. As shown in Figure 4A,B, larger median deformability is observed for MCF10A with the knockout of TP53 (TP53 KO), with ~ 1.24 compared to 1.15 for their isogenic wildtype counterpart. The cell morphology also changed from epithelial (MCF10A TP53 wt) to mesenchymal feature (MCF10A TP53 KO) through the reduced cell-cell adhesion and taking a more elongated shape on the culture plate (Figure 4C).

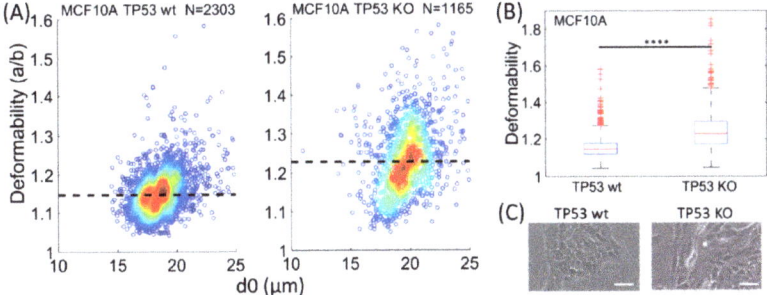

Figure 4. Differential deformability between MCF10A TP53 wildtype (wt) and MCF10A with the knockout of TP53 gene (KO) from the microfluidic hydrodynamic stretching. (**A**) Density scatter plot of the deformability for MCF10A TP53 wt and MCF10A TP53 KO. The dashed lines indicate the median deformability. (**B**) Boxplot and statistical analysis of the above deformability. **** $p < 0.0001$ from two-tailed student t test. (**C**) Bright field images for the cell morphology under the same culture condition. The scale bars denote 20 µm.

The above results suggest that the deletion of the tumor suppressor TP53 gene increases the cellular deformability of MCF10A, which may enhance their invasive potential. As in RKO and HCT116 cell lines, this correlates with cellular morphology changes reminiscent of epithelial-to-mesenchymal transition. Our findings are consistent with a previous study reporting that the deletion of p53 tumor protein that is encoded by the TP53 gene in MCF-10A cells can lead to EMT [9].

3.3. Correlation of Cellular Deformability with Pharmaceutic Drug Treatment

Here we demonstrate the advantage of using hydrodynamic stretching combined with the microsieve method for probing the differential cellular deformability of A549 (adenocarcinomic human alveolar basal epithelial cells) cell lines non-treated and treated with cytochalasin D and docetaxel (a chemotherapy drug).

The microsieve measurements were performed with three replicates, for each of which a syringe loaded with cell mixtures of four different conditions (prelabeled with different fluorescent cell trackers) were used to infuse cells through the microsieve. Both an aliquot of the original cell mixtures before passing to the microsieve and the collected cells trapped on the microsieve (retrieval) were counted using florescence flow cytometry. The relative cell flexibility for the microsieve method is defined as:

Relative cell flexibility = the distribution percentage of cell condition j in the total input cell mixtures/the distribution percentage of cell condition j in the total retrieval cell mixture.

As shown in Figure 5A, the relative cell flexibility of cytochalasin D-treated cells is significantly greater than that of the non-treated cells. Furthermore, the relative cell flexibility increases with the concentration of cytochalasin D used. A significantly decreased relative cell flexibility is observed for docetaxel-treated cells. It is also unlikely that an increasing dosage would lead to further decrease of the relative cell flexibility. In Figure 5B, we also characterize the cell size before and after treatment.

This figure reveals that the mean size of cells treated with cytochalasin D is similar to that of the non-treated ones, whereas a significant increase of cell size is observed after treatment with docetaxel.

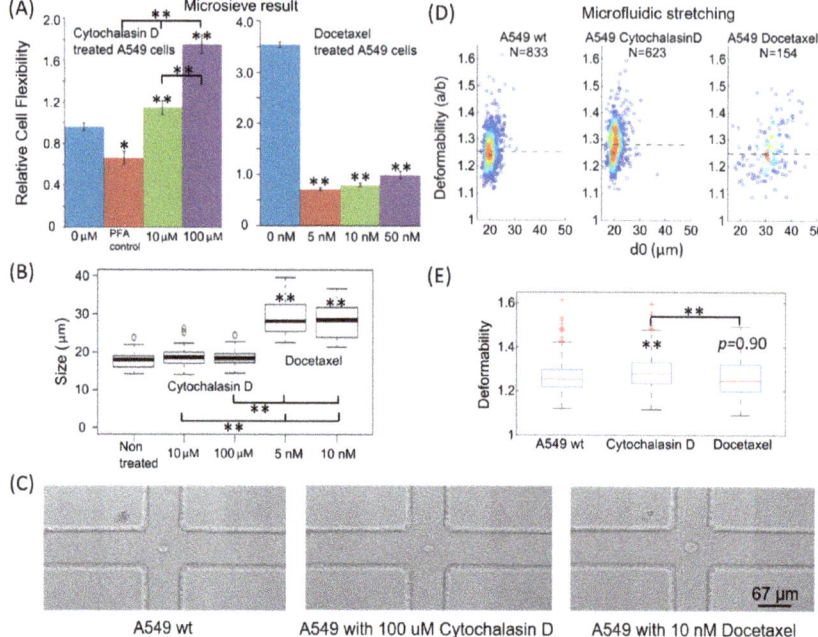

Figure 5. Correlation of cellular deformability of A549 cells with pharmaceutic drug treatment. One-way ANOVA test is performed first to see whether one or more treatment groups are significantly different, followed by post-hoc Tukey HSD test to identify which of the pairs of treatments are significantly different from each other. If not specified, the comparisons are between A549 control (non-treated/wt) and drug treated group, * $p < 0.05$, ** $p < 0.01$. (**A**) Relative cell flexibility from microsieve measurements for A549 control (0 μM), treated with PFA, cytochalasin D (10, 100 μM) and docetaxel (5, 10, 50 nM). (**B**) Statistical analysis for the cell size of A549 non-treated and treated with cytochalasin D (10, 100 μM) and docetaxel (5, 10 nM). Compared to the non-treated group, there is significant increase of the cell size for both 5 and 10 nM docetaxel-treated groups, while no significant change of cell size is found for cytochalasin D-treated groups. (**C**) Representative images of individual cells passing through the center of the stretching region in hydrodynamic stretching microchannel for non-treated A549 (wt) and A549 treated with 100 μM cytochalasin D and 10 nM docetaxel. (**D**) Measured deformability from the microfluidic hydrodynamic stretching for the non-treated A549 and A549 treated with 100 μM cytochalasin D and 10 nM docetaxel: density scatter plot of the deformability. The dashed lines indicate the median deformability. (**E**) Boxplot and statistical analysis of the deformability of (D). A549 treated with cytochalasin D has a significantly higher deformability compared to untreated A549 cells. There is no statistically significant difference between the deformability of docetaxel-treated A549 and the wildtype, $p = 0.90$.

Since both the cell size and deformability will affect the results of the microsieve experiments, it is unclear that whether docetaxel can stiffen the A549 cells. Therefore, we used microfluidic hydrodynamic stretching to compare the deformability of non-treated and treated cells. The results are shown in Figure 5C–E. The median deformability of cytochalasin D-treated A549 cells is significantly larger than that of the non-treated cells, whereas similar deformability is observed between docetaxel-treated and non-treated cells. It can also be seen that the size distribution of docetaxel-treated cells is much wider than the non-treated cells, and the averaged cell diameter is larger. Similarly, the deformability distribution is also wider compared to the non-treated cells, as seen in the boxplot in Figure 5E.

The results from hydrodynamic stretching indicate that the decreased relative cell flexibility for docetaxel-treated cells in the microsieve experiments is probably mainly due to the increased cell size. The increase of cell size is probably because cells get stuck just before mitosis in the cell cycle when treated with docetaxel as docetaxel is a well-established anti-mitotic chemotherapy agent that functions by interfering with cell division [30,31]. Although the drug is thought to induce cellular stiffness in clinics by 'paralyzing' the cytoskeleton, it may not increase cellular stiffness. In summary, hydrodynamic stretching and automated analysis is a feasible method to discriminate between cell size and deformability for the effect of a specific cancer-treatment drug.

4. Discussion

To investigate the correlation between cellular deformability and phenotypes related to EMT, two types of human colon carcinoma cell lines, HCT116 and RKO, were chosen. Recent studies show that RKO cells primarily exhibit mesenchymal features (e.g., elongated morphology, low expression of E-cadherin and high expression of N-cadherin), and have high invasion potential [25,26], whereas HCT116 are reported to demonstrate epithelial features [29]. We also chose MCF10A TP53 wt and its TP53 knockout counterpart, which showed a change of cellular morphology reminiscent of epithelial-to-mesenchymal transition, consistent with a previous study reporting that TP53 deletion in MCF-10A cells can lead to EMT [9]. Using hydrodynamic stretching for these two different sets of cell lines, we find that cells with epithelial-like morphology are stiffer than cells with mesenchymal-like features, supporting the hypothesis that epithelial-to-mesenchymal transition (EMT) may be associated with increased cellular deformability. This is in line with our hypothesis that cellular deformability is responsible, at least in part, for the increased invasive property of mesenchymal cancer cells.

Our results demonstrate that the MCF10A cell line with deletion (knockout) of the tumor suppressor TP53 gene is more flexible compared to its isogenic wildtype counterpart (with the median deformability ~1.24 compared to 1.15), indicating that a single genetic alteration can cause downstream change of cellular biophysical properties relevant to the invasive phenotype of cancer. This result is consistent with a previous study reporting that hENT1 knockdown could induce EMT and reduce cellular stiffness in pancreatic cancer cells [7]. To our knowledge, our study is the first to report a link between the TP53 gene and cellular deformability.

By combining microfluidic hydrodynamic stretching with microsieve cell separation, we find the chemotherapy drug docetaxel has a more profound effect on the cell size of lung carcinoma cell line A549 than on its cellular deformability. This facilitates understanding of the cell separation mechanism of a microsieve for chemotherapy drug screening, as well as providing insights for choosing the right pore size to separate specific cells from phenotype mixtures.

In conclusion, we have established a new method that combines hydrodynamic stretching with microsieve techniques to evaluate alterations in cellular biomechanical properties (e.g., cell size and deformability) that result from cells undergoing EMT and chemotherapy drug treatment. Our results indicate cellular biomechanical properties not only could potentially serve as tools to investigate fundamental properties of cancer cells but also may have the potential to study the progression of cancer invasiveness and the efficiency of chemotherapy, e.g., through measurement of cellular deformability from single cell suspensions obtained from tissue biopsies, pleural effusions or circulating tumor cells.

In the future it would be interesting to include invasion experiments with cells that exhibited higher flexibility after undergoing EMT or chemotherapy drug treatment. Here, we suggest for example invasion experiments of MCF10A TP53 KO and A549 treated with docetaxel vs. their isogenic wildtype counterpart or the untreated group, respectively. A second path for future research could be numerical modeling of the flow field and cell deformation, thus obtaining quantitative values of the stresses exerted on the cells within the elongational flow. This may allow the mechanical properties such as shear elastic modulus, (non-linear) elasticity, or averaged viscosity of the cell content of single cells from different phenotypes to be characterized simultaneously [32,33].

Author Contributions: Conceptualization, F.L., I.C., M.-H.T. and C.D.O.; methodology, F.L., I.C., M.-H.T. and C.D.O.; software, F.L., and C.D.O.; validation, F.L., I.C., M.-H.T. and C.D.O.; formal analysis, F.L., I.C., and J.H.V.; investigation, F.L., I.C., and J.H.V.; resources, F.L., I.C., M.-H.T. and C.D.O.; data curation, F.L., I.C., and J.H.V.; writing—original draft preparation, F.L.; writing—review and editing, I.C., M.-H.T. and C.D.O.; visualization, F.L., I.C., and J.H.V.; supervision, M.-H.T. and C.D.O.; project administration, M.-H.T. and C.D.O.; funding acquisition, M.-H.T. and C.D.O. All authors have read and agreed to the published version of the manuscript.

Funding: We acknowledge financial support with the Nanyang Technological University Research Scholarship through Nanyang Technological University Singapore and The Ministry of Education Singapore.

Acknowledgments: We would like to acknowledge Chon U Chan for the technical support on the design and fabrication of microfluidic hydrodynamic stretching. We also thank Jackie Ying's laboratory and Institute of Bioengineering and Nanotechnology Singapore for the fabrication of the microsieves.

Conflicts of Interest: The authors declare no conflicts of interest.

References

1. Chaffer, C.L.; Weinberg, R.A. A Perspective on Cancer Cell Metastasis. *Science* **2011**, *331*, 1559–1564. [CrossRef]
2. Mani, S.A.; Guo, W.; Liao, M.-J.; Eaton, E.N.; Ayyanan, A.; Zhou, A.Y.; Brooks, M.; Reinhard, F.; Zhang, C.C.; Shipitsin, M.; et al. The epithelial-mesenchymal transition generates cells with properties of stem cells. *Cell* **2008**, *133*, 704–715. [CrossRef] [PubMed]
3. Morel, A.-P.; Lièvre, M.; Thomas, C.; Hinkal, G.; Ansieau, S.; Puisieux, A. Generation of breast cancer stem cells through epithelial-mesenchymal transition. *PLoS ONE* **2008**, *3*, e2888. [CrossRef] [PubMed]
4. Thiery, J.; Acloque, H.; Huang, R.Y.-J.; Nieto, M.A. Epithelial-mesenchymal transitions in development and disease. *Cell* **2009**, *139*, 871–890. [CrossRef] [PubMed]
5. Swaminathan, V.; Mythreye, K.; O'Brien, E.T.; Berchuck, A.; Blobe, G.C.; Superfine, R. Mechanical stiffness grades metastatic potential in patient tumor cells and in cancer cell lines. *Cancer Res.* **2011**, *71*, 5075–5080. [CrossRef]
6. Xu, W.; Mezencev, R.; Kim, B.; Wang, L.; McDonald, J.F.; Sulchek, T.A. Cell Stiffness Is a Biomarker of the Metastatic Potential of Ovarian Cancer Cells. *PLoS ONE* **2012**, *7*, e46609. [CrossRef]
7. Lee, Y.; Koay, E.J.; Zhang, W.; Qin, L.; Kirui, D.K.; Hussain, F.; Shen, H.; Ferrari, M. Human equilibrative nucleoside transporter-1 knockdown tunes cellular mechanics through epithelial-mesenchymal transition in pancreatic cancer cells. *PLoS ONE* **2014**, *9*, e107973. [CrossRef]
8. Bagnall, J.S.; Byun, S.; Begum, S.; Miyamoto, D.T.; Hecht, V.C.; Maheswaran, S.; Stott, S.L.; Toner, M.; Hynes, R.O.; Manalis, S.R. Deformability of Tumor Cells versus Blood Cells. *Sci. Rep.* **2015**, *5*, 18542. [CrossRef]
9. Zhang, Y.; Yan, W.; Chen, X. Mutant p53 disrupts MCF-10A cell polarity in three-dimensional culture via epithelial-to-mesenchymal transitions. *J. Boil. Chem.* **2011**, *286*, 16218–16228. [CrossRef]
10. Di Carlo, D. A mechanical biomarker of cell state in medicine. *J. Lab. Autom.* **2012**, *17*, 32–42. [CrossRef]
11. Sharma, S.; Santiskulvong, C.; Bentolila, L.A.; Rao, J.-Y.; Dorigo, O.; Gimzewski, J.K. Correlative nanomechanical profiling with super-resolution F-actin imaging reveals novel insights into mechanisms of cisplatin resistance in ovarian cancer cells. *Nanomedicine* **2012**, *8*, 757–766. [CrossRef] [PubMed]
12. Puig-De-Morales-Marinkovic, M.; Turner, K.T.; Butler, J.P.; Fredberg, J.J.; Suresh, S. Viscoelasticity of the human red blood cell. *Am. J. Physiol. Physiol.* **2007**, *293*, C597–C605. [CrossRef] [PubMed]
13. Guck, J.; Ananthakrishnan, R.; Mahmood, H.; Moon, T.J.; Cunningham, C.C.; Käs, J. The optical stretcher: A novel laser tool to micromanipulate cells. *Biophys. J.* **2001**, *81*, 767–784. [CrossRef]
14. Sen, S.; Subramanian, S.; Discher, D.E. Indentation and adhesive probing of a cell membrane with AFM: Theoretical model and experiments. *Biophys. J.* **2005**, *89*, 3203–3213. [CrossRef]
15. Rand, R.P.; Burton, A.C. Mechanical Properties of the Red Cell Membrane. I. Membrane Stiffness and Intracellular Pressure. *Biophys. J.* **1964**, *4*, 115–135.
16. Abkarian, M.; Faivre, M.; Stone, H.A. High-speed microfluidic differential manometer for cellular-scale hydrodynamics. *Proc. Natl. Acad. Sci. USA* **2006**, *103*, 538–542. [CrossRef]

17. Li, F.; Chan, C.U.; Ohl, C.-D. Yield strength of human erythrocyte membranes to impulsive stretching. *Biophys. J.* **2013**, *105*, 872–879. [CrossRef]
18. Li, F.; Chan, C.U.; Ohl, C.-D. Rebuttal to a comment by Richard E. Waugh on our article "Yield strength of human erythrocyte membranes to impulsive stretching". *Biophys. J.* **2014**, *106*, 1832–1833. [CrossRef]
19. Quinto-Su, P.A.; Kuss, C.; Preiser, P.R.; Ohl, C.-D. Red blood cell rheology using single controlled laser-induced cavitation bubbles. *Lab Chip* **2011**, *11*, 672–678. [CrossRef]
20. Mohammadzadeh, M.; Li, F.; Ohl, C.-D. Shearing flow from transient bubble oscillations in narrow gaps. *Phys. Rev. Fluids* **2017**, *2*, 014301. [CrossRef]
21. Otto, O.; Rosendahl, P.; Mietke, A.; Golfier, S.; Herold, C.; Klaue, D.; Girardo, S.; Pagliara, S.; Ekpenyong, A.E.; Jacobi, A.; et al. Real-time deformability cytometry: On-the-fly cell mechanical phenotyping. *Nat. Methods* **2015**, *12*, 199–202. [CrossRef] [PubMed]
22. Deng, Y.; Davis, S.P.; Yang, F.; Paulsen, K.; Kumar, M.; Sinnott Devaux, R.; Wang, X.; Conklin, D.; A Oberai, A.; Herschkowitz, J.I.; et al. Inertial Microfluidic Cell Stretcher (iMCS): Fully Automated, High-Throughput, and Near Real-Time Cell Mechanotyping. *Small* **2017**, *13*, 1700705. [CrossRef] [PubMed]
23. Gossett, D.R.; Tse, H.T.K.; Lee, S.A.; Ying, Y.; Lindgren, A.G.; Yang, O.O.; Rao, J.; Clark, A.T.; Di Carlo, D. Hydrodynamic stretching of single cells for large population mechanical phenotyping. *Proc. Natl. Acad. Sci. USA* **2012**, *109*, 7630–7635. [CrossRef] [PubMed]
24. Lim, L.S.; Hu, M.; Huang, M.C.; Cheong, W.C.; Gan, A.T.L.; Looi, X.L.; Leong, S.M.; Koay, E.S.-C.; Li, M.-H. Microsieve lab-chip device for rapid enumeration and fluorescence in situ hybridization of circulating tumor cells. *Lab Chip* **2012**, *12*, 4388–4396. [CrossRef] [PubMed]
25. Turano, M.; Costabile, V.; Cerasuolo, A.; Duraturo, F.; Liccardo, R.; DelRio, P.; Pace, U.; Rega, D.; Dodaro, C.A.; Milone, M.; et al. Characterisation of mesenchymal colon tumour-derived cells in tumourspheres as a model for colorectal cancer progression. *Int. J. Oncol.* **2018**, *53*, 2379–2396. [CrossRef]
26. Feng, B.; Dong, T.T.; Wang, L.L.; Zhou, H.M.; Zhao, H.C.; Dong, F.; Zheng, M.H. Colorectal cancer migration and invasion initiated by microRNA-106a. *PLoS ONE* **2012**, *7*, e43452. [CrossRef]
27. Cima, I.; Kong, S.L.; Sengupta, D.; Tan, I.B.; Phyo, W.M.; Lee, D.; Hu, M.; Iliescu, C.; Alexander, I.; Goh, W.L.; et al. Tumor-derived circulating endothelial cell clusters in colorectal cancer. *Sci. Transl. Med.* **2016**, *8*, 345ra89. [CrossRef] [PubMed]
28. Xu, C.; Prince, J. Snakes, shapes, and gradient vector flow. *IEEE Trans. Image Process.* **1998**, *7*, 359–369.
29. Hur, K.; Toiyama, Y.; Takahashi, M.; Balaguer, F.; Nagasaka, T.; Koike, J.; Hemmi, H.; Koi, M.; Boland, C.R.; Goel, A. MicroRNA-200c modulates epithelial-to-mesenchymal transition (EMT) in human colorectal cancer metastasis. *Gut* **2012**, *62*, 1315–1326. [CrossRef]
30. McKeage, K. Docetaxel. *Drugs* **2012**, *72*, 1559–1577. [CrossRef]
31. Wang, T.; Zhan, Q.; Peng, X.; Qiu, Z.; Zhao, T. CCL2 influences the sensitivity of lung cancer A549 cells to docetaxel. *Oncol. Lett.* **2018**, *16*, 1267–1274. [CrossRef] [PubMed]
32. De Loubens, C.; Deschamps, J.; Georgelin, M.; Charrier, A.; Edwards-Lévy, F.; Leonetti, M. Mechanical characterization of cross-linked serum albumin microcapsules. *Soft Matter* **2014**, *10*, 4561. [CrossRef] [PubMed]
33. De Loubens, C.; Deschamps, J.; Boedec, G.; Leonetti, M. Stretching of capsules in an elongation flow, a route to constitutive law. *J. Fluid Mech.* **2015**, *767*, R3. [CrossRef]

 © 2020 by the authors. Licensee MDPI, Basel, Switzerland. This article is an open access article distributed under the terms and conditions of the Creative Commons Attribution (CC BY) license (http://creativecommons.org/licenses/by/4.0/).

Article

Single-Cell Mechanophenotyping in Microfluidics to Evaluate Behavior of U87 Glioma Cells

Esra Sengul [1] and Meltem Elitas [1,2,*]

[1] Faculty of Engineering and Natural Sciences, Sabanci University, 34956 Istanbul, Turkey; sengulesra@sabanciuniv.edu
[2] Nanotechnology Research and Application Center, Sabanci University, 34956 Istanbul, Turkey
* Correspondence: melitas@sabanciuniv.edu; Tel.: +90-538-810-2930

Received: 13 August 2020; Accepted: 10 September 2020; Published: 11 September 2020

Abstract: Integration of microfabricated, single-cell resolution and traditional, population-level biological assays will be the future of modern techniques in biology that will enroll in the evolution of biology into a precision scientific discipline. In this study, we developed a microfabricated cell culture platform to investigate the indirect influence of macrophages on glioma cell behavior. We quantified proliferation, morphology, motility, migration, and deformation properties of glioma cells at single-cell level and compared these results with population-level data. Our results showed that glioma cells obtained slightly slower proliferation, higher motility, and extremely significant deformation capability when cultured with 50% regular growth medium and 50% macrophage-depleted medium. When the expression levels of E-cadherin and Vimentin proteins were measured, it was verified that observed mechanophenotypic alterations in glioma cells were not due to epithelium to mesenchymal transition. Our results were consistent with previously reported enormous heterogeneity of U87 glioma cell line. Herein, for the first time, we quantified the change of deformation indexes of U87 glioma cells using microfluidic devices for single-cells analysis.

Keywords: glioblastoma; mechanophenotyping; deformation; migration; Label-free; single-cell

1. Introduction

Glioblastoma or Glioblastoma multiforme (GBM), is the most devastating type of brain cancer with nonspecific signs and symptoms and very limited treatment strategies. It results in death in 15 months following diagnosis [1,2]. One of the main challenges for successful treatment of gliomas is its multiform structure [3,4]. Glioma multiforme is highly heterogeneous with different cell types and their complex interactions. Hence, it is an extremely dynamic, hierarchical microenvironment [5]. Moreover, highly activated epithelial-mesenchymal transition (EMT) and existence of glioma stem-like cells (GSCs) drive resistance and relapse in therapy [6–8]. Tremendously increased aggressiveness and invasiveness of GBM requires adequate single-cell tools to avoid masking rare cells and to allow quantifying heterogeneity according to morphologic, phenotypic, and functional properties of glioma cells in a population.

Since Scherer investigated distinct morphological patterns of infiltrating glioma cells in 1940, mechanophenotyping of glioma cells has been researched to identify selective biomarkers or diagnostic indicators for the incidence and prognosis of GBM [9,10]. Mechanical properties of cells, being closely linked to homeostasis of cells and their own microenvironment, have been hallmarks for defining healthy and malignant conditions of cells particularly in metastatic cancer [11–14]. Several different technologies have been developed to measure mechanical properties of cells including flow cytometry [15–17], atomic force microscopy (AFM) [18,19], magnetic twisting cytometry (MTC) [20], parallel-plate rheometry [11,21], optical stretching (OS) [22], optical tweezer [23,24], microfluidic ektacytometry [25,26], and micropipette aspiration [27,28]. Utilizing these modern tools for delineating mechanophenotypic properties of cells including stiffness, adhesiveness, viscosity, deformation (ratio of the area to volume), morphology, and migration trajectories of cells have been extensively investigated in cancer biology [29–31].

Mechanistic studies of glioma cells, emphasis on cell morphology, migration, proliferation, and invasion have been mostly interrogated in microfabricated 3D-growth chambers confined with gels that mimic the extracellular matrix (ECM). Microfabricated cell culture platforms have provided a well-controlled microenvironment for glioma cells while allowing high-resolution time-lapse microscopy imaging [32]. In addition, cell-cell adhesion strength and cell-deformability properties of glioma cells have been investigated by AFM and single cell force spectroscopy methods [33]. A pioneering work by Kaufman's research group revealed collagen concentration-dependent growth and motility patterns of GBM cells using bulk rheology, phase-contrast, confocal reflectance, and CARS microscopy techniques [34]. Along this study, Ulrich et al. reported that ECM rigidity increased glioma motility, proliferation, and spread using time-lapse and immunofluorescence experiments [34]. Furthermore, Memmel et al. applied scanning electron microscopy (SEM) and single cell electrorotation techniques with traditional methods to characterize cell surface morphology and membrane folding of GBM cell lines [35,36].

In this study, we have investigated biomechanical properties of U87-MG (HTB-14 ™) glioma cells creating two glioma cell culturing conditions; i-U87 category: U87 glioma cells fed by the regular growth medium, and ii-U87-C category: U87 glioma cells were cultured in 50% growth medium supplemented by the 50% macrophage-depleted medium. We evaluated the behavior of glioma cells under these two growth conditions using both population-based traditional assays and microfluidic-based single-cell analysis. To characterize biomechanical heterogeneity of glioma U87 cell line at single-cell level, we measured proliferation, migration, motility, and deformation of glioma cells. We determined the expression levels of E-cadherin and Vimentin proteins between the U87 and U87-C populations according to immunostaining assays using fluorescence microscopy images.

2. Materials and Methods

2.1. Cell Culture

The U87 MG (HTB-14™) human glioma and the U937 (CRL 1593.2™) human histiocytic lymphoma monocyte cell line was purchased from ATCC (American Type Culture Collection). Cells were grown in 75 cm^2 cell culturing flasks and 6-well plates (TTP, Switzerland at 37 °C with 5% CO_2 (NUVE, Turkey). The U87 cells were cultured in the DMEM medium (Dulbecco's modified Eagle's medium, Roswell Park Memorial Institute), 10% fetal bovine serum (FBS/Sigma-Aldrich, St. Louis, MO), 1% Penicillin/Streptomycin (Pen/Strep, Pan Biotech, Germany). The U937 monocytes were maintained in RPMI 1640 medium (Roswell Park Memorial Institute, Pan Biotech, Germany), 10% FBS (Sigma-Aldrich, St. Louis, MO, USA). The human histiocytic lymphoma macrophages were differentiated from the U937 monocytes through stimulation of 3×10^5 cells/mL in 5 mL RPMI 1640, 10% FBS with 5 µL working solution of 10% Phorbol 12-myristate 13-acetate (PMA, Pan Biotech, Germany) obtained from 10 ng/mL PMA/dimethyl sulfoxide (DMSO, Pan Biotech, Germany) stock solution according to standard protocols for macrophage differentiation.

The conditioned medium was harvested from U937-differentiated macrophages grown in RPMI 1640, 10% FBS for 72 h at 37 °C with 5% CO_2. The collected media was centrifuged at 3000 rpm for 5 min on a centrifuge (Hettich-EBA-20, Germany) and filtered through a 0.2-µm filter (GVS Filter Technology, UK) and then the harvested supernatant is freshly used in the experiments.

2.2. Microfluidic Chip Fabrication

Microfluidic chips were designed using the CleWin 4.0 layout editor. The microfluidic cell culture platform has one inlet and outlet for cell loading and medium feeding, Figure 1. The microchamber allows cells to be cultured and visualized (1280 × 500 µm, h = 50 µm) with two types of pillars. We designed the circular (r = 90 µm, h = 50 µm) and trapezoid pillars (a = 80 µm, b = 215, h = 50 µm) to prevent polydimethylsiloxane (PDMS) from collapsing inside the cell culture microchamber and observe whether different pillar geometries effect migration of the cells. Their dimensions and pitches were determined to deform flow, slow down cells, and distribute them randomly in the cell culture microchamber [37]. To observe the imaging area by 10× objective, the distance between the pillars were determined to be 390 and 190 µm for the circular and trapezoidal ones, respectively.

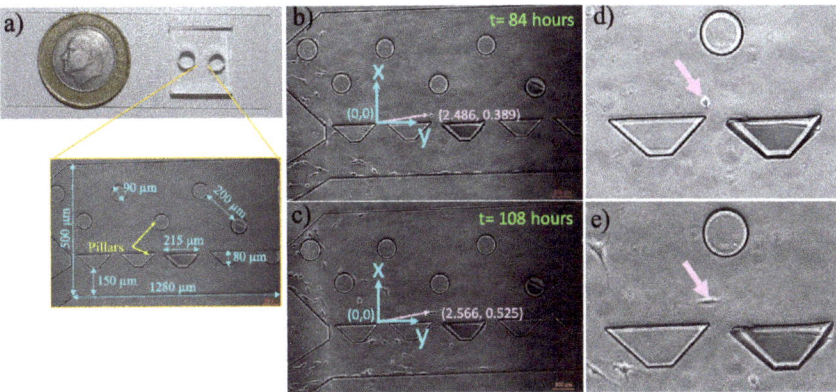

Figure 1. Microfluidic cell culture platform and measurement of single-cell migration in the microfluidic device. (**a**) The polydimethylsiloxane (PDMS) microfluidic chamber on a glass slide, (**b**) the micrographs of microfluidic chamber with pointed pillars and dimensions. (**c**) The blue line indicates the position of the coordinate system (0, 0) with *x*- and *y*-axis. The pink arrow points to the position of a cell according to origin, (**b**) x: 2.486, y: 0389 at 84 h, (**c**) x: 2.566, y: 0.525 at 108 h, (**d**) and (**e**) demonstrate the zoomed images of this cells in (**b**) and (**c**), respectively. The scale bar shows 100 µm.

The designs were patterned on a thin film chromium deposited photomask (Cr-blank) using a Vistec/EBPG5000plusES Electron Beam Lithography System. SU-8 2025 (SU-8® 2025, MicroChem) was spin-coated on a four inches silicon wafer to obtain 50-µm tall structures. Next, the photoresist-coated wafers were soft baked (65 °C, 3 min and 95 °C, 5 min) and exposed to UV light (160 mJ cm^{-2}, Midas/MDA-60MS mask aligner). Upon two consecutive post-baking processes (65 °C, 1 min and 95 °C, 5 min), SU-8 was developed (MicroChem's SU-8 developer). The microfluidic chips were obtained using elastomeric polymer PDMS (Sylgard® 184, Dow Corning, Midland, MI, USA) [38]. Five-mm biopsy punchers (Robbins Instruments, Chatham, MA, USA) were used to create inlet and outlet ports. The PDMS chips were irreversibly bonded both inside a 6-well plate and on a glass slide using a Corona system (BD20-AC, Electro-Technic Products Inc., Chicago, IL, USA), Figure 1.

2.3. Microfluidic Chip Preparation and Culturing Cells in the Microfluidic Device

To prepare the microfluidic chip, all reagents and microchips were placed into the incubator (37 °C and 5% CO_2, NUVE, Turkey) for 30 min. Prior to cell loading, to eliminate air bubbles inside the microfluidic culture chamber, the warm medium was added using a 200-µL micropipette (Corning, New York, NY, USA). U87 glioma cells were grown as explained above (2.1 Cell culture), trypsinized (Pan Biotech, Germany), and resuspended in the DMEM medium to obtain 1.6 × 10^5 cells/mL. Next, cell suspension with 8 × 10^4 cells/mL was injected into the microfluidic device. Afterwards, the U87 glioma chips were mounted in the incubator (37 °C and 5% CO_2, NUVE, Turkey) and their medium replaced with the fresh 40-µL DMEM medium every 24 h. The U87 glioma conditional (U87-C) chips were generated by replacing the regular DMEM medium with the conditioned medium when the cells were grown in the DMEM medium (37 °C and 5% CO_2, NUVE, Turkey) overnight. The conditioned medium (explained in 2.1 Cell Culture) within microfluidic devices was also regularly refreshed once a day. Each experiment was independently performed in triplicate.

2.4. Cell Growth in a 12-Well Cell Culture Plate

Once the U87 cells reached 75 to 85% of confluency, cells were trypsinized (Pan Biotech, Germany) and resuspended in the fresh DMEM medium with trypan blue dye (Sigma-Aldrich, Darmstadt, Germany). Next, the number of viable cells was counted using a hemocytometer (Marienfeld, Germany). U87 cells were seeded at a density of 1×10^5 cells/well in a 12-well cell culture plate (TPP, Switzerland) and allowed to adhere overnight in the incubator (37 °C and 5%, NUVE, Turkey). After overnight incubation, three wells were assigned as U87 and fed by the regular DMEM medium, while the others were assigned as U87-C and fed by the conditioned medium for five days in the incubator (37 °C and 5%, NUVE, Turkey). Both medium replacements and cell count determinations were performed for 24 h for five days. To determine the cell numbers, the cells were trypsinized (Pan Biotech, Germany), centrifuged (Eba 20, Hettich, Germany) at 1800 rpm for 10 min. The cell pellets were collected and suspended in the fresh medium with trypan blue dye. Total viable cells were counted using a hemocytometer (Marienfeld, Germany) for both U87 and U87-C growth conditions, Figure 2. Each experiment was independently performed in duplicate. Results were represented by means ± standard errors.

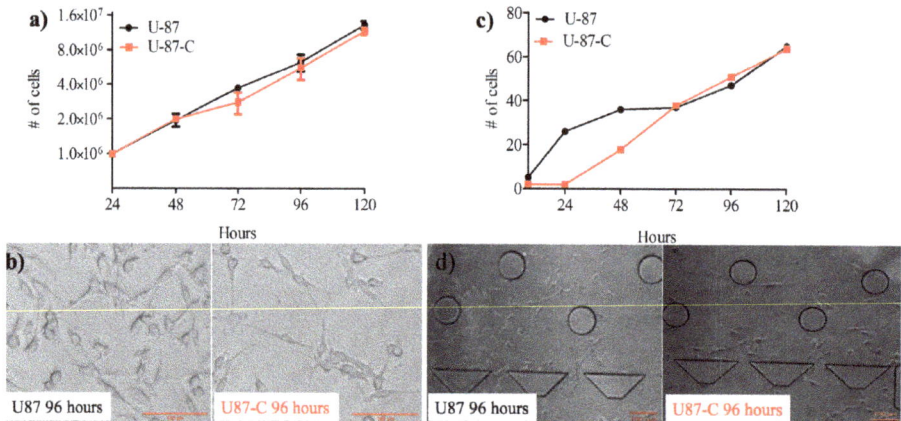

Figure 2. Growth comparison. Glioma cells were grown in DMEM medium (U87) and in 50% DMEM and 50% macrophage-depleted medium (U87-C) in the 6-well culture dish and microfluidic platform. (**a**) The number of viable cells for five days in a 6-well plate, (**b**) the micrographs of U87 and U87-C cells for a 96-h growth. The scale bar shows 20 µm. (**c**) The number of viable cells for five days in the microfluidic device, (**d**) micrographs of glioma cells for the 96-h growth. The scale bar shows 100 µm. The number of cells present the mean ± standard error for two independent experiments.

2.5. Cell Migration by Wound Healing in a 12-Well Cell Culture Plate

The wound healing assay was performed in a 6-well cell culture plate (TPP, Switzerland) where U87 cells were seeded at a density of 1×10^6 cells/mL using a 2 mL DMEM medium. The prepared culture plate was kept in the incubator at 37 °C, 5% CO_2, and allowed cells to adapt their microenvironment and adhere to the surface of the 6-well plate. Next, six wells of U87 culture were grown in the regular medium while the other six wells were maintained in the conditioned medium (Section 2.1 Cell Culture chapter) until the cells became confluent (two days). Before the scratch wound was created in the cell monolayer using a 200-µL-pipette tip (Eppendorf, Germany), phase-contrast images of the cells were observed using an inverted fluorescent microscope, the Zeiss Axio Observer (Carl Zeiss Axio Observer Z1, Germany) equipped with a 10× objective and the AxioCam Mrc5 camera. The wells were washed with the medium to remove the floating cells, 3 mL of either regular or conditioned medium per well were added into the wells, the images of the wells were acquired.

Upon 24 h of incubation, 10 µM DAPI (Life Sciences 33342) and 10 µM Propidium iodide solution (PI, Sigma-Aldrich P4864) were added into the wells containing 3 mL of medium. Next, the same microscope setting was used to acquire 24-h images of the cells to quantify the number of cells that migrated towards the wound area. The exposure rates of DAPI, FITC, and PI channels were 100, 600, and 400 mS, respectively. The excitation and emission values were 495/519 nm, long pass (LP) 515 nm for FITC, 535/617 nm LP 590 nm for PI, and 358–410 nm, LP 420 nm for DAPI, respectively. This experiment was independently performed in duplicate. The wound closure analysis was performed using ImageJ (Version 2.0 National Institutes of Health, Rockville, MD, USA). The number of migrated cells into the scratched region was manually counted using the ImageJ software. The unpaired, two-tailed Student's t-test was performed to determine whether the migrated-cell number difference between the U87 and U87-C populations was significant using the GraphPad Prism 5 software.

2.6. Measurement of Single-Cell Migration in the Microfluidic Device

U87 cells were harvested as explained in 2.1 Cell Culture. Two separate microfluidic devices were prepared either with the regular medium or conditioned medium (as explained in 2.3 Microfluidic chip preparation and culturing cells in the microfluidic device). Cells were not stained by fluorescent dyes to eliminate the effects of labeling on biomechanical properties of cells. Next, the microfluidic chips were maintained at 37 °C, 5% CO_2 (NUVE, Turkey). The microfluidic chips that were prepared with the regular medium were replenished with the regular medium, while the conditional medium was used to refill the conditional microfluidic devices. Microfluidic platforms were daily monitored on the microscope and images of the microchamber area were acquired using the Zeiss Axio Observer Z1 inverted microscope equipped with a 10× lens and an AxioCam Mrc5 camera. Upon imaging, the medium was replenished inside the microchambers, and microfluidic platforms were placed into the cell culture incubator. Using the obtained images, the migration distances of the cells were determined, Figure 2. The single cell migration assays in the microfluidic devices were performed for five days.

To consistently measure the migration distances of the cells, the coordinate system was defined as shown in Figure 1. The positions of the cells were defined with respect to the origin of the coordinate system. Therefore, the movement of the cells can be quantified between the inlet and outlet ports inside the microfluidic cell culture chamber for five days. Positions of single cells were manually measured using the ImageJ software (version 2.0). The velocities of cells were also calculated.

2.7. Immunohistochemistry

U87 cells were maintained in DMEM and U87-C glioma cells were grown in a 50% growth and 50% macrophage-depleted medium with the density of 50,000 cells/well on a round coverslip inside the 12-well plates. The cells were fixed using 4% Paraformaldehyde (Boster BioSciences, Cat No: AR1068) at room temperature for 30 min. Next, cells were permeabilized using 0.1% Triton-X100 (Sigma, T8787) and 0.1% Bovine Serum Albumin (BSA, Sigma A2058) in PBS. Upon PBS washes, coverslips were incubated overnight with primary antibody against E-cadherin (Abcam: ab1416) and Vimentin (Abcam: ab8978). Both primary antibody concentrations were adjusted to 1:100 in 2.5% BSA and 0.05% Triton X-100. The secondary antibody Alexa Fluor 488 (Thermo Fisher #A10680) was used at 1:200. DAPI (Life Sciences 33342) staining was performed after incubation. Coverslips were mounted using 50% glycerol in 1X PBS at room temperature. The coverslips were observed by a plan-apochromat 63×/1.42 Oil objective lens of a fluorescent microscope (Olympus BX60, Japan) using U-MNU2 filter with 365/10 nm excitation and 420 LP nm emission values. The exposure rates of DAPI and FITC channels were 133 µs and 175.3 µs, respectively.

The expression levels of the E-cadherin and Vimentin proteins were quantified using ImageJ (version 2.0 National Institutes of Health, Rockville, MD, USA) and GraphPad Prism 5. The number of the cells was counted using the DAPI-stained nuclei of the cells using ImageJ. Next, the total region of the acquired image area was defined to measure total fluorescence intensity from Alexa Fluor 488 dye using the ImageJ software. Afterwards, obtained row intensity density values were divided by the number of the cells. We used one-way ANOVA Tukey's comparison test to determine whether the expression levels of E-cadherin and Vimentin proteins were significant between the U87 and U87-C glioma groups (GraphPad Prism 5).

3. Results

3.1. Influence of Conditioned Medium on U87 Proliferation in 12-Well Plate and Microfluidic Device

Two sets of glioma cells were prepared to be cultured. One of them was labeled as U87, it was grown in the regular growth medium. The other one was marked as U87-C that was cultured in the 50% growth and 50% macrophage-depleted medium (conditioned medium). Both U87 and U87-C cultures were maintained in 12-well plates for five days (see Material and Methods, Sections 2.3 and 2.4).

We first examined the growth differences between U87 and U87-C glioma cells in culture dishes. We counted viable cells using a hemocytometer according to trypan blue staining. Figure 3a,b shows that the cell viability within the conditioned medium was similar to the growth medium (Figure 3c,d), there was no significant growth difference for 120 h according to the unpaired t-test ($p = 0.407$). Hence, this result validated that the phenotypic differences between U87 and U87-C were able to be further studied.

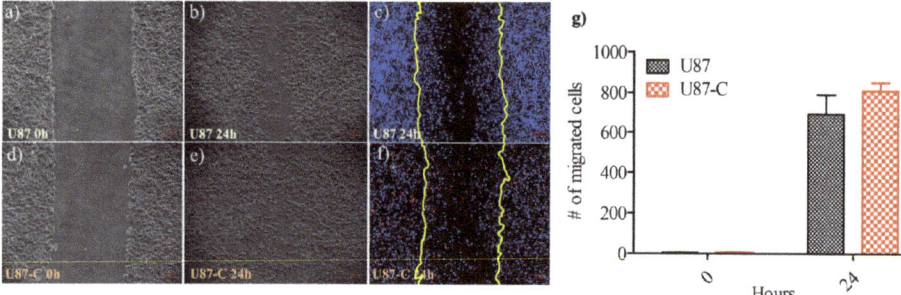

Figure 3. Analysis of U87 and U-7-C cells migration by the in vitro wound-healing assay. The phase images of U87 cells when the (**a**) wound created at 0 h, (**b**) phase images of wound closure at 24 h, (**c**) fluorescence images of wound closure at 24 h, the nucleus of the cells are labeled with DAPI and displayed in blue, dead cells are PI-stained and shown in red, yellow lines present the wound area created at 0 h. The same settings were applied for U87-C (**d**–**f**). The images acquired with 10× magnification; the scale bar shows 100 µm. (**g**) The number of migrated cells at 0 and 24 h. The results represent the mean ± standard deviation of two independent experiments. There was no significant difference according to the Student's unpaired, two-tailed t-test, $p = 0.9051$.

The microfabricated cell culture chamber allows proliferation and migration of the cells from the inlet port to the outlet port while enabling the monitoring behavior of cells at single-cell resolution, Figure 1. Initially, the culture medium was injected from the inlet port through the microchamber to the outlet port. The culture medium was a growth medium and conditioned medium for U87 and U87-C, respectively. Next, bubbles were removed from the microchamber, and then cell suspension was added into the inlet port. The gravity-driven laminar flow owing to 400-µm height difference between the inlet port and cell culture microchannel was obtained for gentle nutrient delivery and waste removal through the microchamber. Hence, the fluid flow did not mechanically interrupt the behavior of the cells. Figure 2c displays the number of viable cells for normal and conditioned growth environments in the microfluidic chips. As shown in Figure 2d, images of the microchannel were acquired every 24 h and the number of viable cells were manually counted. The viability was defined according to cell division via following single cells. A label-free analysis was performed to eliminate staining-induced phenotype variations.

3.2. Influence of Conditional Medium on U87 Cell Migration by Wound Healing Assay

The wound healing assay was performed in a 12-well cell culture plate (see Material and Methods, Section 2.5). The six wells of U87 culture were grown in the regular medium while the other six wells were maintained in the conditioned medium. The scratch wound was created in the cell monolayer using a 200-µL-pipette tip. The phase-contrast images of the wells were acquired immediately after scratching the cell monolayer and 24 h later. Next, the number of migrated cells into the scratch area was manually counted using the ImageJ software, Figure 3a–f. Each experiment was independently performed in duplicate. The U87-C group showed increased migration compared with the U87 category, however, the difference was not significant according to the Student's t-test (p = 0.9051), Figure 3g.

Moreover, when U87 cells were cultured in the regular medium, the cells proliferated and remained on the borders of the scratched region instead of migrating to the cell-free regions. However, it was not observed for the U87-C group where U87 cells were grown in 50% DMEM and 50% macrophage depleted RPMI. Therefore, the crowdedness of the glioma cells in the central wound area was higher for the U87-C category, Figure 3.

3.3. Influence of Conditional Medium on U87 Cell Migration Using a Microfluidic Device

We examined the movement of the cells both in the regular medium (U87) and conditional medium (U87-C) between 48 to 120 h in a microfluidic cell culture chamber. The first 48 h of cell culture in a microfluidic device were used to allow cells to adhere to a glass surface and proliferate. Next, the images of the cells were acquired for every 12 h and analyzed as illustrated in Figure 1 (see Material and Methods, Section 2.6). Twenty cells from DMEM cultured and 20 cells from conditional- medium cultured glioma cells were selected, and their positions were recorded at 12-h intervals.

Figure 4 presents the changes of positions on the x-y axis for U87 (Figure 4a) and U87-C (Figure 4d) groups. Migration of the cells was demonstrated on the x-axes (Figure 4b,e) and on the y-axes (Figures 4c and 5f). The migration distances of the U87 cells were shorter in comparison to U87-C. Movement of the U87 cell population was more uniform than the U87-C cell population on y-axis.

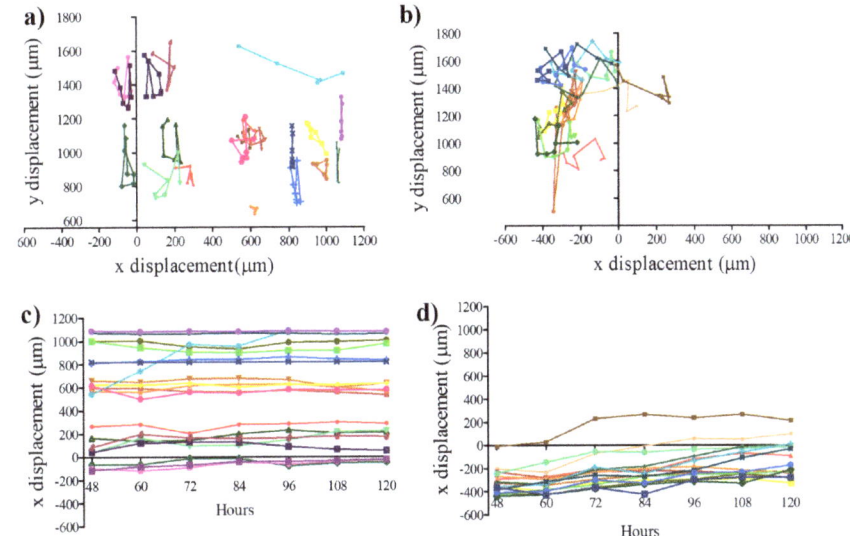

Figure 4. Cont.

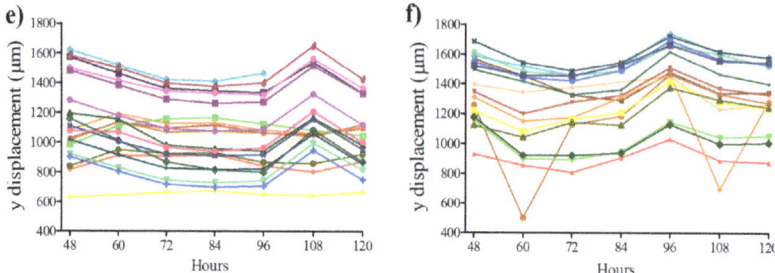

Figure 4. Migration of single cells in the microfluidic cell culture chamber. Coordinates of the cells in the microfluidic cell culture device were measured every 12 h between 48 to 120 h. Movement of the U87 cells (**a**) at x-y axes (**b**) on the x-axis, (**c**) on the y-axis. Movement of the U87-C cells (**d**) at x-y axes, (**e**) on the x-axis, (**f**) on the y-axis. The number of analyzed cells for each group is 20. U87 indicates that cells cultured in DMEM medium, U87-C defines that cells were grown in 50% DMEM and 50% macrophage-used RPMI medium. Each color represents the single cells and color coding was consistent in each group.

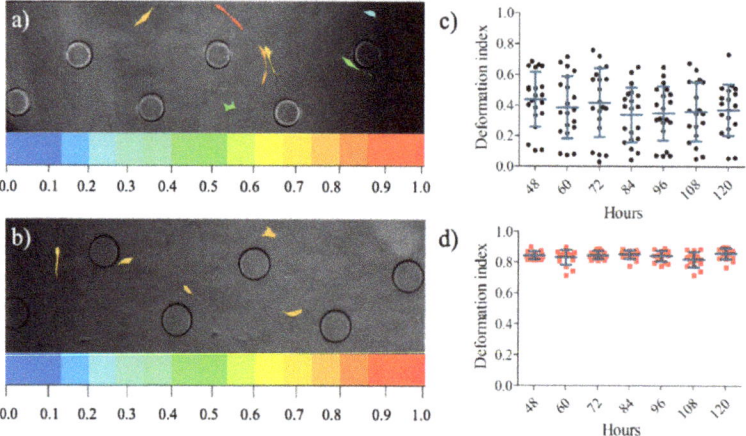

Figure 5. Deformation indexes of the cells in the microfluidic device. The phase images of the cells with a colorimetric deformation scale, the range of deformability from coolest colors (blue: 0) to warm colors (red:1) represent enhanced deformability indexed (**a**) U87 population, (**b**) U87-C population. The deformation indexes of 20 glioma cells between 48- and 120-h (**c**) for U87 population, (**d**) for U87-C population. The data presents the mean ± standard error.

3.4. Influence of Conditional Medium on U87 Cell Migration Using a Microfluidic Device

Upon assessing the positions of the single cells in the microfluidic device, the area and perimeter measurements of 20 cells from U87 and U87-C populations were performed. The images of the cells were obtained every 12 h between 48 to 120 h. ImageJ was used to manually measure the diameter and perimeter of the cells. The single-cell deformation indexes (D) of each cell were calculated using Equation (1) [39,40], where π is 3.14. Figure 5 illustrates the deformation index of single cells in the microfluidic cell culture platform.

$$D = 1 - \frac{2\sqrt{\pi \, \text{Area}}}{\text{Perimeter}} \tag{1}$$

Figure 5 elucidates that deformation indexes of glioma cells in regular medium were more heterogeneous (Figure 6a,c) in comparison to glioma cells cultured in conditioned medium (Figure 5b,d). The deformability difference between these two populations was significant according to the Student's two-tailed t-test, $p < 0.0001$, Figure 6a. Figure 6b shows that the area to perimeter ratio of U87 population is greater than U87-C populations according to the Student's two-tailed t-test, $p < 0.0258$.

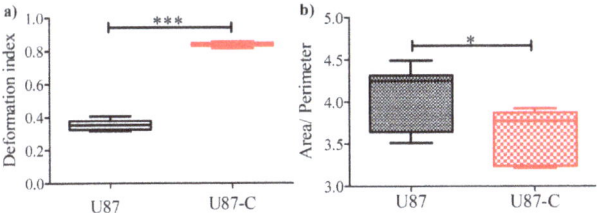

Figure 6. Comparisons of deformation indexes and area/perimeter. (**a**) Deformation indexes differences between U87 and U87-C. Student's two-tailed *t*-test was applied, $p < 0.0001$. (**b**) Differences of the area to perimeter ratio according to the Student's two-tailed t-test, p-value is 0.0258. * and *** implies for $p < 0.05$ and $p < 0.0001$, respectively.

3.5. Influence of Conditional Medium on the Expression of E-cadherin and Vimentin

E-cadherin and Vimentin proteins are among the molecular markers of epithelium-to-mesenchymal transition (EMT) [41]. Since, the U87-C glioma population gained more deformation (Figure 5) and migratory (Figure 6) properties in comparison with U87 glioma cells according to single-cell analysis, we evaluated the expression levels of E-cadherin and Vimentin proteins in a 12-well plate using immunostaining, Figure 7a,b. The expression levels of E-cadherin and Vimentin proteins were not significantly different between U87 and U87-C glioma cells on day 3. The weak expression level of E-cadherin significantly decreased both for U87 ($p < 0.05$) and U87-C ($p < 0.001$) populations from 3 to 5 days. Figure 7c,d displays the expression level of Vimentin protein, which was moderately weak for U87-C glioma cells and weak for U87 cells. When the expression level of Vimentin was compared between day 3 and 5, the decrease in the expression level of Vimentin was not significant for U87 glioma cells. In contrast, the decrease was significant for the U87-C glioma population ($p < 0.001$).

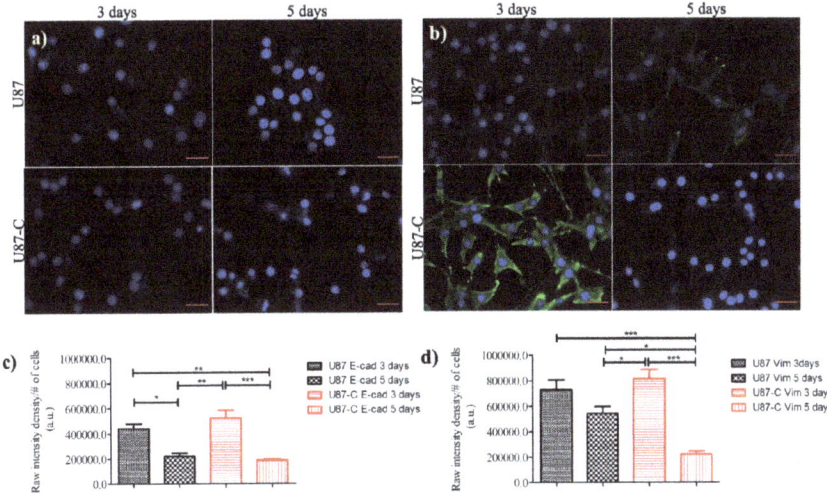

Figure 7. Comparisons of E-cadherin and Vimentin expressions. (**a**) Immunofluorescence staining of E-cadherin (E-cad), (**b**) Vimentin (Vim) proteins (green) with nuclei counterstained (blue) by DAPI. The scale bar is 25 µm, the magnification is x63. Quantification of (**c**) E-cadherin and (**d**) Vimentin expressions of U87 and U87-C glioma cells for 3 and 5 days. The one-way analysis of variance Tukey's multiple comparison test was applied. *, **, and *** denotes for $p < 0.05, 0.01$, and $p < 0.0001$, respectively.

4. Discussion

The complex, dynamic, and highly heterogeneous microenvironment of glioblastoma tumors present a chicken and egg problem when we focus on understanding the interactions of glioma and immune cells. Therefore, we need to develop new methods and tools to discover important, measurable properties of cells that are not adequately measurable using traditional macroscale techniques. Elucidation of the mechanisms underlying the heterogeneity of tumor microenvironment requires quantification of cellular properties at single-cell level for a large number of cell populations and compiling the obtained results with the existing data in the literature. However, still there is a gap to be bridged between well-established, macroscale types of assay and microfluidic-based, microscale methods that match to the length and time scales of cells [42–44]. In this study, our aim was to investigate the influence of macrophage-secreted proteins on the behavior of glioma cells when glioma and macrophage cells were not directly in contact with each other, while integrating traditional bulk assays and microfluidic single-cell platforms.

First, we examined whether there was a significant growth difference between the U87 and U87-C glioma populations using both traditional cell culture dishes and microfabricated cell culture platforms. Figure 2 showed that there were not significant growth differences between U87 and U87-C, which allowed us to further investigate the influence of conditioned medium on biomechanical properties of U87 glioma cells. Since, the observed mechanophenotyipc differences of U87 cells might be due to macrophage-secreted proteins in the conditioned medium, not owing to growth deficiency. Although the growth of U87 cells were similar in both U87 and U87-C conditions, the micrographs of cells in Figure 2b,d showed that both in 12-well plate and microfabricated cell culture platform glioma cells were more elongated in the U87-C population. We observed that the number of rounded cells were higher for the U87 glioma population within the microfluidic chamber, where the cells adhered on the glass surface. Hence, the U87 glioma cell morphology is also dependent on the substrate stiffness, as previously reported for the LN229, LN18, and LBC3 glioma cell lines and glioma primary cells [45–48]. Moreover, our results agreed with the previous research that revealed morphological heterogeneity of U87 glioma cell line [49].

We next performed the wound healing assay using a 6-well plate. Our results presented that there was no significant difference between the number of migrated glioma cells for U87 and U87-C culture conditions ($p = 0.9051$), Figure 3g. As an important difference between U87 and U87-C glioma populations, glioma cells were distributed on the scratched region of the wound in the U87-C population compared to the U87 group where cells were more adhered to the leading edges of the wound.

To further investigate the influence of conditioned medium on the U87 cell migration, we quantified the behavior of glioma cells by cell tracking analysis. We measured the displacement of glioma cells in the microfluidic platform both for U87 and U87-C populations, Figure 4. Our results verified that glioma cells moved longer distances with a relatively high migration speed in the U87-C population, when cells were fed by the conditioned medium, Figure 4a–d. Both U87 and U87-C cell populations disseminated more on the y-axis. Movement of the cells along the x-axis were more uniformly distributed. Herein, neither U87 nor U87-C populations exhibited directionality in their movement. Still, these results show that GBM cells exhibited high heterogeneity in migration displacement, orientation, and velocity [49,50]. To the best of our knowledge, our study presents for the first-time single cell tracking analysis of glioma cells for 120 h in the microfluidic platform. Mostly, migration and morphology assays have been performed for shorter time frames (0–10 h) in vitro assays [35,47].

Afterwards, we assessed whether indirect contact between glioma and macrophage cells influences deformation capability of glioma cells. We determined deformation indexes (DI) of glioma cells according to area and perimeter measurements of glioma cells. Figure 5 shows that glioma cells in the U87-C population have significantly higher deformation indexes (DI > 0.8) in comparison to the U87 glioma population (DI < 0.4), $p < 0.0001$. Additionally, glioma cells displayed higher deformation heterogeneity in the regular growth medium in comparison to the conditioned medium.

Taken together, culturing glioma cells in 50% DMEM and 50% macrophage-depleted medium influenced morphology, motility, and deformation of glioma cells. To evaluate whether these altered mechanical phenotypes were linked to epithelium-to-mesenchymal transition, we assessed the expression levels of E-cadherin and Vimentin proteins both in U87 and U87-C populations. E-cadherin and Vimentin proteins are among the molecular markers of epithelium-to-mesenchymal transition (EMT) [41]. The EMT process introduces a malignant phenotype, spindle-shaped morphology, and metastatic functions for cancer cells by altering the activation of transcription factors, expression levels of specific microRNAs and cell-surface proteins, as well as organization of cytoskeletal proteins. Epithelial cells become mesenchymal with low levels of E-cadherin and high levels of vimentin expressions. Our results indicated that the indirect effect of macrophages under our experimental conditions did not provide mesenchymal phenotype to U87 glioma cells.

5. Conclusions

This study demonstrated a single-cell mechanophenotyping approach for U87 glioma cells while integrating microfluidic cell culture platforms with macroscale traditional assays. We cultured glioma cells either in the regular growth medium, denoted as U87 cell population, or in the 50% regular growth medium and 50% macrophage-depleted medium (conditioned medium), referred to as U87-C. We cultured both U87 and U87-C glioma populations in traditional cell culture dishes and microfluidic platforms for five days. We quantified proliferation, morphology, motility, migration, and deformation properties of glioma cells using single-cell analysis, which are directly linked to biomechanical features of cells. Our results presented that there was no significant growth difference between U87 and U87-C glioma populations, however U87-C glioma cells exhibited a slightly weaker proliferation in the microfluidic device. U87 and U87-C glioma populations were morphologically heterogeneous both in the bulk and microfluidic assays. We observed that the macrophage-conditioned stimulation provided glioma cells with slightly increased motility and extremely significant deformation capabilities ($p < 0.0001$). We measured the expression levels of E-cadherin and Vimentin proteins to assess whether the phenotype of glioma cells in the U87-C population were transformed into the mesenchymal phenotype. However, immunostaining experiments verified that observed phenotypic changes were statistically E-Cadherin and Vimentin independent. Our results confirmed enormous heterogeneity of U87 glioma cell line in terms of mechanophenotypic properties. Herein, we integrated microscale and macroscale growth conditions and quantified mechanophenotypic properties of glioma cells thanks to the microfluidic cell culture platform. Considering our results, integration of microfabricated, single-cell level and traditional, population-level assays will be the future of modern techniques in cell biology.

Author Contributions: Conceptualization, M.E.; methodology, M.E. and E.S.; validation, M.E. and E.S.; formal analysis, E.S.; investigation, M.E. and E.S.; resources, M.E.; writing—original draft preparation, E.S.; writing—review and editing, M.E.; visualization, M.E. and E.S.; supervision, M.E.; project administration, M.E., funding acquisition, M.E. All authors have read and agreed to the published version of the manuscript.

Funding: This research was funded by the Scientific and Technological Research Council of Turkey (TUBITAK), grant number, 217S616.

Conflicts of Interest: The authors declare no conflict of interest.

References

1. Ostrom, Q.T.; Gittleman, H.; Liao, P.; Rouse, C.; Chen, Y.; Dowling, J.; Wolinsky, Y.; Kruchko, C.; Barnholtz-Sloan, J. CBTRUS Statistical Report: Primary Brain and Central Nervous System Tumors Diagnosed in the United States in 2007-2011. *Neuro-Oncol.* **2014**, *16* (Suppl. S4), iv1–iv63. [CrossRef] [PubMed]
2. Komori, T. The 2016 WHO Classification of Tumours of the Central Nervous System: The Major Points of Revision. *Neurol. Med.-Chir.* **2017**, *57*, 301–311. [CrossRef]
3. Holland, E.C. Glioblastoma Multiforme: The Terminator. *Proc. Natl. Acad. Sci. USA* **2000**, *97*, 6242–6244. [CrossRef] [PubMed]

4. Gimple, R.C.; Kidwell, R.L.; Kim, L.J.; Sun, T.; Gromovsky, A.D.; Wu, Q.; Wolf, M.; Lv, D.; Bhargava, S.; Jiang, L.; et al. Glioma Stem Cell–Specific Superenhancer Promotes Polyunsaturated Fatty-Acid Synthesis to Support EGFR Signaling. *Cancer Discov.* **2019**, *9*, 1248–1267. [CrossRef] [PubMed]
5. Schiffer, D.; Annovazzi, L.; Casalone, C.; Corona, C.; Mellai, M. Glioblastoma: Microenvironment and Niche Concept. *Cancers* **2018**, *11*, 5. [CrossRef] [PubMed]
6. Takashima, Y.; Kawaguchi, A.; Yamanaka, R. Promising Prognosis Marker Candidates on the Status of Epithelial–Mesenchymal Transition and Glioma Stem Cells in Glioblastoma. *Cells* **2019**, *8*, 1312. [CrossRef]
7. Phillips, H.S.; Kharbanda, S.; Chen, R.; Forrest, W.F.; Soriano, R.H.; Wu, T.D.; Misra, A.; Nigro, J.M.; Colman, H.; Soroceanu, L.; et al. Molecular Subclasses of High-Grade Glioma Predict Prognosis, Delineate a Pattern of Disease Progression, and Resemble Stages in Neurogenesis. *Cancer Cell* **2006**, *9*, 157–173. [CrossRef]
8. Guardia, G.D.A.; Correa, B.R.; Araujo, P.R.; Qiao, M.; Burns, S.; Penalva, L.O.F.; Galante, P.A.F. Proneural and Mesenchymal Glioma Stem Cells Display Major Differences in Splicing and LncRNA Profiles. *npj Genom. Med.* **2020**, *5*. [CrossRef]
9. Scherer, H.J. A Critical Review: The Pathology of Cerebral Gliomas. *J. Neurol. Neurosur. Psychiatry* **1940**, *3*, 147–177. [CrossRef]
10. Kozminsky, M.; Sohn, L.L. The Promise of Single-Cell Mechanophenotyping for Clinical Applications. *Biomicrofluidics* **2020**, *14*, 031301. [CrossRef]
11. Wu, P.-H.; Aroush, D.R.-B.; Asnacios, A.; Chen, W.-C.; Dokukin, M.E.; Doss, B.L.; Durand-Smet, P.; Ekpenyong, A.; Guck, J.; Guz, N.V.; et al. A Comparison of Methods to Assess Cell Mechanical Properties. *Nat. Methods* **2018**, *15*, 491–498. [CrossRef] [PubMed]
12. Wirtz, D.; Konstantopoulos, K.; Searson, P.C. The Physics of Cancer: The Role of Physical Interactions and Mechanical Forces in Metastasis. *Nat. Rev. Cancer* **2011**, *11*, 512–522. [CrossRef] [PubMed]
13. Guck, J.; Schinkinger, S.; Lincoln, B.; Wottawah, F.; Ebert, S.; Romeyke, M.; Lenz, D.; Erickson, H.M.; Ananthakrishnan, R.; Mitchell, D.; et al. Optical Deformability as an Inherent Cell Marker for Testing Malignant Transformation and Metastatic Competence. *Biophys. J.* **2005**, *88*, 3689–3698. [CrossRef] [PubMed]
14. Tse, H.T.; Gosset, R.D.; Moon, Y.S.; Masaeli, M.; Sohsman, M.; Ying, Y.; Mislick, K.; Adams, P.; Rao, J.; Di Carlo, D. Quantitative Diagnosis of Malignant Pleural Effusions by Single-Cell Mechanophenotyping. *Sci. Transit. Med.* **2013**, *5*, 212ra163. [CrossRef] [PubMed]
15. Urbanska, M.; Muñoz, H.E.; Bagnall, J.S.; Otto, O.; Manalis, S.R.; Carlo, D.D.; Guck, J. A Comparison of Microfluidic Methods for High-Throughput Cell Deformability Measurements. *Nat. Methods* **2020**, *17*, 587–593. [CrossRef]
16. Eluru, G.; Srinivasan, R.; Gorthi, S.S. Deformability Measurement of Single-Cells at High-Throughput With Imaging Flow Cytometry. *J. Lightwave. Technol.* **2015**, *33*, 3475–3480. [CrossRef]
17. Krutzik, P.O.; Nolan, G.P. Fluorescent Cell Barcoding in Flow Cytometry Allows High-Throughput Drug Screening and Signaling Profiling. *Nat. Methods* **2006**, *3*, 361–368. [CrossRef]
18. Pelling, A.E.; Veraitch, F.S.; Chu, C.P.-K.; Mason, C.; Horton, M.A. Mechanical Dynamics of Single Cells during Early Apoptosis. *Cell Motil. Cytoskeleton.* **2009**, *66*, 409–422. [CrossRef]
19. Radmacher, M. Studying the Mechanics of Cellular Processes by Atomic Force Microscopy. *Methods Cell Biol.* **2007**, 347–372. [CrossRef]
20. Puig-De-Morales, M.; Grabulosa, M.; Alcaraz, J.; Mullol, J.; Maksym, G.N.; Fredberg, J.J.; Navajas, D. Measurement of Cell Microrheology by Magnetic Twisting Cytometry with Frequency Domain Demodulation. *J. Appl. Physiol.* **2001**, *91*, 1152–1159. [CrossRef]
21. Thoumine, O.; Ott, A.; Cardoso, O.; Meister, J.-J. Microplates: A New Tool for Manipulation and Mechanical Perturbation of Individual Cells. *J. Biochem. Bioph. Meth.* **1999**, *39*, 47–62. [CrossRef]
22. Guck, J.; Ananthakrishnan, R.; Mahmood, H.; Moon, T.J.; Cunningham, C.C.; Käs, J. The Optical Stretcher: A Novel Laser Tool to Micromanipulate Cells. *Biophys. J.* **2001**, *81*, 767–784. [CrossRef]
23. Huang, N.-T.; Zhang, H.-L.; Chung, M.-T.; Seo, J.H.; Kurabayashi, K. Recent Advancements in Optofluidics-Based Single-Cell Analysis: Optical on-Chip Cellular Manipulation, Treatment, and Property Detection. *Lab. Chip.* **2014**, *14*, 1230–1245. [CrossRef] [PubMed]
24. Musielak, M. Red Blood Cell-Deformability Measurement: Review of Techniques. *Clin. Hemorheol. Microcirc.* **2009**, *42*, 47–64. [CrossRef] [PubMed]
25. Artmann, G. Microscopic Photometric Quantification of Stiffness and Relaxation Time of Red Blood Cells in a Flow Chamber. *Biorheology* **1995**, *32*, 553–570. [CrossRef] [PubMed]

26. Scott, M.D.; Matthews, K.; Ma, H. Assessing the Vascular Deformability of Erythrocytes and Leukocytes: From Micropipettes to Microfluidics. In *Current and Future Aspects of Nanomedicine*; IntechOpen: London, UK, 2019. [CrossRef]
27. Zhao, R.; Sider, K.L.; Simmons, C.A. Measurement of Layer-Specific Mechanical Properties in Multilayered Biomaterials by Micropipette Aspiration. *Acta Biomater.* **2011**, *7*, 1220–1227. [CrossRef]
28. Lee, L.M.; Lee, J.W.; Chase, D.; Gebrezgiabhier, D.; Liu, A.P. Development of an Advanced Microfluidic Micropipette Aspiration Device for Single Cell Mechanics Studies. *Biomicrofluidics* **2016**, *10*, 054105. [CrossRef]
29. Tee, S.-Y.; Bausch, A.R.; Janmey, P.A. The Mechanical Cell. *Curr. Biol.* **2009**, *19*. [CrossRef]
30. Leggett, S.E.; Patel, M.; Valentin, T.M.; Gamboa, L.; Khoo, A.S.; Williams, E.K.; Franck, C.; Wong, I.Y. Mechanophenotyping of 3D Multicellular Clusters Using Displacement Arrays of Rendered Tractions. *Proc. Natl. Acad. Sci. USA* **2020**, *117*, 5655–5663. [CrossRef]
31. Shah, M.K.; Garcia-Pak, I.H.; Darling, E.M. Influence of Inherent Mechanophenotype on Competitive Cellular Adherence. *Ann. Biomed. Eng.* **2017**, *45*, 2036–2047. [CrossRef]
32. Diao, W.; Tong, X.; Yang, C.; Zhang, F.; Bao, C.; Chen, H.; Liu, L.; Li, M.; Ye, F.; Fan, Q.; et al. Behaviors of Glioblastoma Cells in in Vitro Microenvironments. *Sci. Rep.* **2019**, *9*. [CrossRef] [PubMed]
33. Andolfi, L.; Bourkoula, E.; Migliorini, E.; Palma, A.; Pucer, A.; Skrap, M.; Scoles, G.; Beltrami, A.P.; Cesselli, D.; Lazzarino, M. Investigation of Adhesion and Mechanical Properties of Human Glioma Cells by Single Cell Force Spectroscopy and Atomic Force Microscopy. *PLoS ONE* **2014**, *9*, e112582. [CrossRef] [PubMed]
34. Kaufman, L.; Brangwynne, C.; Kasza, K.; Filippidi, E.; Gordon, V.; Deisboeck, T.; Weitz, D. Glioma Expansion in Collagen I Matrices: Analyzing Collagen Concentration-Dependent Growth and Motility Patterns. *Biophys. J.* **2005**, *89*, 635–650. [CrossRef] [PubMed]
35. Ulrich, T.A.; Pardo, E.M.D.J.; Kumar, S. The Mechanical Rigidity of the Extracellular Matrix Regulates the Structure, Motility, and Proliferation of Glioma Cells. *Cancer Res.* **2009**, *69*, 4167–4174. [CrossRef] [PubMed]
36. Memmel, S.; Sukhorukov, V.L.; Höring, M.; Westerling, K.; Fiedler, V.; Katzer, A.; Krohne, G.; Flentje, M.; Djuzenova, C.S. Cell Surface Area and Membrane Folding in Glioblastoma Cell Lines Differing in PTEN and p53 Status. *PLoS ONE* **2014**, *9*, e87052. [CrossRef]
37. Texier, B.D.; Laurent, P.; Stoukatch, S.; Dorbolo, S. Wicking through a confined micropillary array. *Microfluid. Nanofluid.* **2016**, *20*, 53. [CrossRef]
38. Elitas, M.; Sadeghi, S.; Karamahmutoglu, H.; Gozuacik, D.; Turhal, N.S. Microfabricated platforms to quantitatively investigate cellular behavior under the influence of chemical gradients. *Biomed. Phys. Eng. Express* **2017**, *3*, 03023. [CrossRef]
39. Nawas, A.A.; Ubanska, M.; Herbig, M.; Nötzel, M.; Kräter, M.; Rosendahk, P.; Herold, C.; Toepfner, N.; Kubánková, M.; Goswami, R.; et al. Intelligent Image-based Deformation Assisted Cell Sorting with Molecular Specificity. *Nat. Methods* **2020**, *17*, 595–599. [CrossRef]
40. Elitas, M.; Sengul, E. Quantifying Heterogeneity According to Deformation of the U937 Monocytes and U937-Differentiated Macrophages Using 3D Carbon Dielectrophoresis in Microfluidics. *Micromachines* **2020**, *11*, 576. [CrossRef]
41. Shankar, J.; Nabi, I.R. Actin Cytoskeleton Regulation of Epithelial Mesenchymal Transition in Metastatic Cancer Cells. *PLoS ONE* **2015**, *10*, e0119954. [CrossRef]
42. Paguirigan, A.L.; Beebe, D.J. Microfluidics Meet Cell Biology: Bridging the Gap by Validation and Application of Microscale Techniques for Cell Biological Assays. *BioEssays* **2008**, *30*, 811–821. [CrossRef] [PubMed]
43. Duncombe, T.A.; Tentori, A.M.; Herr, A.E. Microfluidics: Reframing Biological Enquiry. *Nat. Rev. Mol. Cell Biol.* **2015**, *16*, 5540567. [CrossRef] [PubMed]
44. Chiu, D.T.; deMello, A.J.; Di Carlo, D.; Doyle, P.S.; Hansen, C.; Maceiczyk, R.M.; Wooton, R.C.R. Small but Perfectly Formed? Successes, Challenges, and Opportunities for Microfluidics in the Chemical and Biological Sciences. *Chem* **2017**, *2*, 201–223. [CrossRef]
45. Bangasser, B.L.; Shamsan, G.A.; Chan, C.E.; Opoku, J.N.; Tüzel, E.; Schlichtmann, B.W.; Kasim, J.A.; Fuller, B.J.; McCullough, B.R.; Rosenfeld, S.S.; et al. Shifting the Optimal Stiffness for Cell Migration. *Nat. Commun.* **2017**, *8*, 15313. [CrossRef] [PubMed]
46. Koh, I.; Cha, J.; Park, J.; Choi, J.; Kang, S.-G.; Kim, P. The Mode and Dynamics of Glioblastoma Cell Invasion into a Decellularized Tissue-Derived Extracellular Matrix-Based Three-Dimensional Tumor Model. *Sci. Rep.* **2018**, *8*, 4608. [CrossRef] [PubMed]

47. Pogoda, K.; Bucki, R.; Byfield, F.J.; Cruz, K.; Lee, T.; Marcinkiewicz, C.; Janmey, P.A. Soft Substrates Containing Hyaluronan Mimic the Effects of Increased Stiffness on Morphology, Motility, and Proliferation of Glioma Cells. *Biomacromolecules* **2017**, *18*, 3040–3051. [CrossRef] [PubMed]
48. Manini, L.; Caponnetto, F.; Bartolini, A.; Ius, T.; Mariuzzi, L.; Di Loreto, C.; Beltrami, A.P.; Cesselli, D. Role of Microenvironment in Glioma Invasivness: What We Learned from In Vitro Models. *Int. J. Mol. Sci.* **2018**, *19*, 147. [CrossRef]
49. Fayzullin, A.; Sandberg, C.J.; Spreadbury, M.; Saberniak, B.M.; Grieg, Z.; Skaga, E.; Langmoen, I.A.; Vik-Mo, E.O. Phenotypic and Expressional Heterogeneity in the Invasive Glioma Cells. *Transl. Oncol.* **2019**, *12*, 122–133. [CrossRef]
50. Parker, J.J.; Canoll, P.; Niswander, L.; Kleinschmidt-DeMasters, B.K.; Foshay, K.; Waziri, A. Intratumoral Heterogeneity of Endogenous Tumor Cell Invasive Behavior in Human Glioblastoma. *Sci. Rep.* **2018**, *8*, 18002. [CrossRef]

© 2020 by the authors. Licensee MDPI, Basel, Switzerland. This article is an open access article distributed under the terms and conditions of the Creative Commons Attribution (CC BY) license (http://creativecommons.org/licenses/by/4.0/).

Article

Quantifying Heterogeneity According to Deformation of the U937 Monocytes and U937-Differentiated Macrophages Using 3D Carbon Dielectrophoresis in Microfluidics

Meltem Elitas [1,2,*] and Esra Sengul [1]

1. Faculty of Engineering and Natural Sciences, Sabanci University, Istanbul 34956, Turkey; sengulesra@sabanciuniv.edu
2. Sabanci University Nanotechnology Research and Application Center, Istanbul 34956, Turkey
* Correspondence: melitas@sabanciuniv.edu; Tel.: +90-538-810-2930

Received: 19 May 2020; Accepted: 8 June 2020; Published: 8 June 2020

Abstract: A variety of force fields have thus far been demonstrated to investigate electromechanical properties of cells in a microfluidic platform which, however, are mostly based on fluid shear stress and may potentially cause irreversible cell damage. This work presents dielectric movement and deformation measurements of U937 monocytes and U937-differentiated macrophages in a low conductive medium inside a 3D carbon electrode array. Here, monocytes exhibited a crossover frequency around 150 kHz and presented maximum deformation index at 400 kHz and minimum deformation index at 1 MHz frequencies at 20 $V_{peak-peak}$. Although macrophages were differentiated from monocytes, their crossover frequency was lower than 50 kHz at 10 $V_{peak-peak}$. The change of the deformation index for macrophages was more constant and lower than the monocyte cells. Both dielectric mobility and deformation spectra revealed significant differences between the dielectric responses of U937 monocytes and U937-differentiated macrophages, which share the same origin. This method can be used for label-free, specific, and sensitive single-cell characterization. Besides, damage of the cells by aggressive shear forces can, hence, be eliminated and cells can be used for downstream analysis. Our results showed that dielectric mobility and deformation have a great potential as an electromechanical biomarker to reliably characterize and distinguish differentiated cell populations from their progenitors.

Keywords: dielectrophoresis; deformation; mobility; heterogeneity; macrophage; monocyte

1. Introduction

Dielectric parameters are among the essential biophysical properties of cells and can be associated with various immune and blood diseases [1–5]. Permeability and conductivity of the membrane and cytoplasm define dielectric properties of a cell in a specific microenvironment, which may change due to surface area of the cell as given by its size and shape; expression levels of surface proteins; form of cytoplasm; composition of cytos7ol; the surface charge density of the membrane; the morphologic complexity of membrane surfaces such as ruffles, microvilli, and blebs; as well as due to interfacial polarization of ions at the cell surfaces. Discovery of electrophysiological properties of cells, such as dielectrophoretic mobility, membrane relaxation period, crossover frequency difference, etc., relies on the phenomenon of dielectrophoresis (DEP), described by Herbert Pohl in 1951 [6]. Yet, intensive research has been conducted to utilize dielectrophoretic properties of cells to be label-free biomarkers for immune and blood diseases [7,8]. In this study, we interrogated whether dielectric movement and deformation measurements provide a specific, label-free, sensitive electromechanical biomarker for U937 monocytes and U937-differentiated macrophages.

Monocytes and macrophages can be considered as active machines that can immediately adapt to their microenvironment for pathogenesis and homeostasis through altering their electromechanical properties [9,10]. They are highly heterogenetic cells with their morphology, location, tissue-specific relations, and functional capabilities [11,12]. When examined by electron microscopy, the monocytes are spherical cells and they have microvilli and microcytotic vesicles, hence, their membrane surfaces have several ruffles and blebs, whereas macrophages have an irregular shape with electron-dense membrane-bound lysosomes. Besides, the microenvironment in which macrophages differentiate defines their shape, biochemistry and function [13]. Although we have been still investigating and discovering their new functions, such as the roles of macrophages in the electrical conduction of heart [14], in general, we know that monocytes enroll in tumor formation and invasion via metastasis and angiogenesis [15,16], macrophages are employed in pathogen recognition, phagocytosis [17], removal of dead cells and cellular debris [18] and tissue homeostasis [19,20]. Their diverse functions are continuously controlled by their dynamic microenvironment [21–24].

A pioneering work, sharing the purpose of determining electrical properties of mammalian cells according to their life cycle, was presented by Eisenberg and Doljanski in 1962. They measured the electrokinetic properties of liver cells in growth processes [25]. Next, Dr Petty's research group reported heterogeneous distribution of electrophoretic mobilities of human monocyte subpopulations [26], while Dr Bauer and Dr Hannig determined the changes of the electrophoretic mobility (EM) of human monocytes during in vitro maturation into macrophages [27]. The current research direction, which investigates the change of cellular dielectrophoretic properties during the cell cycle, maturation or differentiation, mostly relies on determining the first crossover frequencies and measuring migration differences of cells [7,8,28,29]. Along the same lines, our previous investigations have presented the dielectrophoretic characterization and separation of U937 monocytes and U937-differentiated macrophages using their crossover frequencies and dielectrophoretic mobility differences according to their membrane permittivity and conductivity in a low conductive DEP buffer [30–32]. However, none of our previous studies have revealed dielectrophoresis-induced mechanical deformation of cells. Similarly, Tonin et al. interrogated electrophoretic mobility (EPM) during yeast growth and observed a nonmonotonic behavior during the cell cycle. They concluded that the maximal EPM occurred at the initial stage of the growth, and it strongly reduced at its final stage [33]. Song et al. employed DEP to sort human mesenchymal stem cells and their differentiation progeny, osteoblasts. Their results showed that osteoblasts experienced stronger DEP forces that laterally migrated them, whereas human mesenchymal stem cells remained on their original trajectories [34]. Dr. Salmanzadeh and his group used contactless DEP and observed that the trapping voltage of mouse ovarian surface epithelial cells increased as the cells progressed from a non-tumorigenic to a tumorigenic phenotype [35].

On the other hand, DEP has been utilized as a tool to stretch cells for characterization of their mechanical properties. It has provided great potential to implement single-cell biomechanical tests with high-throughput, automation, low complexity and cost, high scalability and portability in comparison to conventional biomechanical techniques, such as atomic force microscopy [36], optical tweezers [4,37], magnetic twisting cytometry [38], micropipette aspiration [39], diffraction phase microscopy [40] and microfluidic ektacytometry [41–43]. In this concept, Guido et al. demonstrated the capability of this new technique by characterizing deformability of cancerous MCF7 and noncancerous MCF10A cells [44]. Du and coworkers used this technique to reveal the biophysical properties of healthy, uninfected and infected red blood cells by Plasmodium falciparum malaria parasites [45].

In this study, we utilized dielectrophoresis to study the electromechanical properties of monocytes and macrophages that might quantify their population heterogeneity [11,45,46]. We measured the movement and calculated the deformation indexes of cells [47] under the influences of dielectrophoretic forces when 10–20 $V_{peak\text{-}to\text{-}peak}$ (V_{pp}) voltage with frequencies ranging from 50 kHz to 1 MHz have been applied.

2. Materials and Methods

2.1. DEP Buffer Preparation and Conductivity Measurement

DEP buffer with low electrical conductivity was prepared to keep cells viable during the processes of dielectrophoresis. As it has been previously reported [31], the low conductive DEP buffer [48] was composed of 8.6% sucrose (product no: LC-4469.1, NeoFroxx, Hesse, Germany), 0.3% glucose (CAS number 59-99-7, Sigma-Aldrich, Darmstadt, Germany) and 0.1% bovine serum albumin in distilled water (BSA, product code: P06-1391050, PAN-Biotech, Aidenbach, Germany).

The conductivity of the DEP buffer was 0.002 S/m, as measured by a Corning Model 311 Portable conductivity meter at room temperature (Cambridge Scientific Products, Watertown, MA, USA).

2.2. Cell Culture

In this study, U937 human monocyte cells (ATCC number: CRL1593.2) provided from ATCC (American Type Culture Collection, Manassas Virginia) and U937-differentiated macrophages were obtained by phorbol 12-myristate 13-acetate (PMA, Sigma Aldrich) treatment of U937 monocytes.

U937 cells were maintained in RPMI 1640 medium (Product Number: P04-18047, PAN-Biotech, Aidenbach, Germany) with 10% fetal bovine serum (FBS) (PAN Biotech, catalogue number: P40-37500, Aidenbach, Germany) using a T75 tissue culture flask (TPP® Sigma, catalogue number: Z707554) at 37 °C with 5% CO_2 in humidified air. U937 cells were grown until 80–90% confluency. Cells were centrifuged at 3000 rpm (Z601039, Hettich® EBA 20 centrifuge, Merck, Darmstadt, Germany) for 5 min. The number of cells was determined using a hemocytometer (Marienfeld, Germany). The final cell concentration was adjusted to 3×10^6 cells/mL.

The macrophage differentiation was performed using the 10 ng/mL concentration of the PMA treatment of 3×10^6 U937 cells in 22.1 cm^2 plates (TPP® Product No:93060, Trasadingen, Switzerland) for 72 h. Next, the cells were maintained in medium without PMA for 48 h. Then, the cells were collected by treating with the 0.25% (v/v) Trypsin-EDTA (PAN Biotech, catalogue number: P10-019100, Aidenbach, Germany) solution. The cells were centrifuged at 1800 rpm (Z601039, Hettich® EBA 20 centrifuge, Merck, Darmstadt, Germany) for 10 minutes to remove the remaining culture medium and washed twice using the DEP buffer.

2.3. 3D Carbon DEP Device

The fabrication process and features of the 3D carbon DEP devices were previously reported [48,49]. The carbon electrode array, a 1.8 mm wide, 3.2 cm long channel, was featured 218 intercalated rows with 14 or 15 electrodes each [50,51]. Individual electrodes had a height of 100 µm and a diameter of 50 µm (Figure 1). The numerical analysis to estimate the induced fluidic, electromagnetic and dielectrophoretic forces in the 3D carbon electrode array was earlier studied using both finite element analysis and numerical models [48–51].

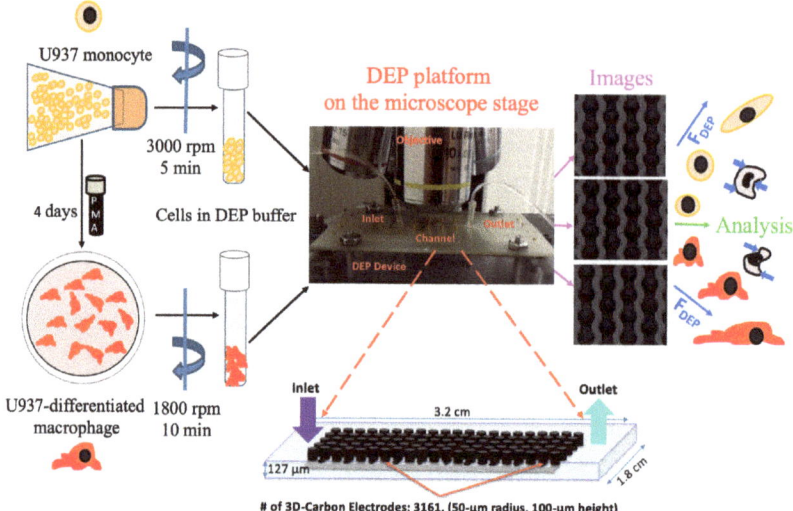

Figure 1. Schematic illustration of the experimental setup comprising the cell preparation step, 3D carbon electrode array, imaging and single-cell analysis.

2.4. Experimental Setup

The experimental setup consisted of a signal generator (Model: GFG-8216A, GW Instek, New Taipei City, Taiwan) with an oscilloscope (Part Number: 54622D, Agilent Technologies, Santa Clara, CA, USA) to create and observe the electric field, a desktop-acquired upright microscope (Model: Nikon ME600 Eclipse, Nikon Instruments Inc., Melville, NY, USA) to monitor cells and acquire images and a programmable syringe pump (Model: NE-1000, New Era Pump Systems Inc., Farmingdale, NY, USA) to flow the cells into the 3D carbon DEP device. We used 20–200 µL pipette tips (Manufacturer ID: 3120000917, Eppendorf, Hamburg, Germany) to connect microperforated Tygon tubing (Manufacturer ID: AAQ02103-CP S-54-HL, Cole-Parmer, Vernon Hills, IL, USA) into the inlet and outlet ports of the 3D carbon-DEP chip (Figure 1).

The experiment started with the sterilization of the electrode array using 70% ethanol and rinsing with deionized (DI) water using a syringe pump with a 20 µL/min flow rate. Next, the microfluidic chip was filled with the DEP buffer and the bubbles were removed. Then, 40 µL of the cell suspension was injected into the chip using a syringe pump with 10 µL/min flow rate. When the cells reached the electrode area, the flow was stopped, and the cells were released for 30 s. The experiments were started when the electric field was applied using the signal with 10–20 V_{pp} frequencies ranging from 50 kHz–1 MHz [30,31].

2.5. Image Acquisition and Data Analysis

The image sequences of cells were recorded using the Nikon ME600 Eclipse upright microscope (Nikon Instruments Inc., Melville, NY, USA) with 10× magnification in tiff sequence format. The VideoLAN Client (VLC, VideoLAN version 1.8, Paris, France) program was used to convert image sequences into the movies.

The acquired images were manually analyzed using open-access ImageJ software (Version 2.0 National Institutes of Health, Rockville, MD, USA). The crossover frequencies of single cells were determined by computing the movement of the cells according to their initial positions, as described in references [30,31]. In total, 50 monocyte cells and 30 macrophage cells were followed, and their positions were recorded. Using GraphPad Prism (Version 5.0) software, Student's t-test was performed to compare dielectric mobilities of monocyte and macrophage populations. * implies that data are significantly different with $p < 0.5$.

The deformation index was calculated by manually measuring the height and width of 45 single monocyte and macrophage cells, and these single cells were continuously monitored, in each frequency. One-way analysis of variance and Tukey's multiple comparison test were carried out using GraphPad Prism (Version 5.0) software to determine the significance. * and ** indicate that data are significantly different with $p < 0.5$ and $p < 0.05$, respectively. All measurements were provided in detail in the figure legends.

3. Results

3.1. Dielectrophoretic Movement

DEP offers the possibility to affect the movement of polarized particles in the non-uniform electric field. We can define the DEP force according to the difference between the dielectric properties of the particle and its suspension medium [52,53].

$$F_{DEP} = 2\pi r^3 \varepsilon_m Re(K(\omega))\nabla E^2 \quad (1)$$

The DEP force (F_{DEP}) is related to the radius of the particle, the permittivity of the surrounding medium (ε_m), the real part of the Clausius–Mossotti factor ($Re(K(\omega))$) and the applied electric field (E). The Clausius–Mossotti factor is defined as given by

$$K(\omega) = \frac{(\varepsilon_c^* - \varepsilon_m^*)}{(\varepsilon_c^* + 2\varepsilon_m^*)} \quad (2)$$

Here, ε_c^* is known as the complex permittivity of a cell and ε_m^* is the complex permittivity of the surrounding medium. The subscripts "m" and "c" mean suspending medium and cells, respectively. The complex permittivity can be expressed as

$$\varepsilon^* = \varepsilon + \frac{j\sigma}{\omega} \quad (3)$$

where ε is the permittivity, σ is the conductivity and ω ($\omega = 2\pi f$) includes the electric field frequency. When the value of the $Re(K(\omega))$ is positive, the particle is attracted by the strong electric field region referred to as positive DEP (pDEP). When the value of the $Re(K(\omega))$ is negative, the particle is repelled by the high electric field region referred to as negative DEP (nDEP). The crossover frequency can be defined as the cessation of the particle motion, which is specific for the particles.

To quantify heterogeneity of monocytes and macrophages according to their dielectrophoretic behaviors, we applied the non-uniform AC electric field and determined the location of the cells in each frequency ranging from 50 kHz to 1 MHz (Figure 1). Our previous work presents the determination of the crossover frequencies in detail for the immune cells [30].

The translational movement of the cells was generated by dielectrophoretic forces and no fluid flow can introduce any drag force on the cells. Figure 2 demonstrates the number of cells that experienced strong pDEP (3), pDEP (2), weak pDEP (1), CF (0), weak nDEP (−1), nDEP (−2), strong nDEP (−3) at 50, 100, 200, 300, 400 and 1000 kHz frequencies when 20 and 10 V_{pp} voltages were applied for monocytes and macrophages, respectively.

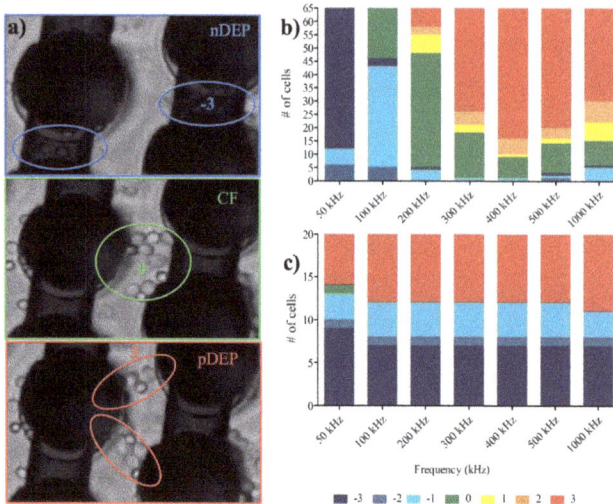

Figure 2. Dielectrophoretic responses of monocytes and macrophages: (**a**) Positions of the cells in the electrode array when they are influenced by nDEP, crossover frequency (CF) and pDEP, respectively; (**b**) Translational movement of U937 monocytes under 20 Vpp, 50 kHz–1 MHz nonuniform AC field; (**c**) Translational movement of U937-differentiated macrophages under 10 Vpp, 50 kHz–1 MHz nonuniform AC field. The cool colors show the number of nDEP- behaved cells due to repelling DEP forces while the warm colors demonstrate pDEP-responded cells owing to attractive DEP forces. Zero means the crossover frequency with zero movements, which is coded in green color. n = 80 for monocytes, n = 30 for macrophages.

Figure 2 demonstrates the dielectrophoretic behavior of the U937 monocytes and U937-differentiated macrophages under the influence of nonuniform electric field within the 3D carbon electrode array. Figure 2b shows that monocyte cells experienced nDEP to pDEP forces with increasing frequencies (n = 80 monocyte cells). The crossover frequencies of monocytes were between 100 to 200 kHz. The uniformity of pDEP responses of the monocytes was improved with increasing frequencies ranging from 200 kHz to 1 MHz, the strongest nDEP (−3, dark blue), the strongest pDEP (3, red) see Supplementary Video 1.

On the other hand, when the same experiment was performed using the U937-differentiated macrophage cells, they mostly exhibited pDEP behavior (warm colors yellow-red colors) and their weak crossover frequency was around 50 kHz (green), as shown in Figure 2b. The fraction of macrophage cells which immediately presented pDEP response was greater than the nDEP subpopulation. The number of nDEP experienced cells were not broadly changed in comparison to monocyte cells. Since most of the macrophage cells immediately experienced pDEP behavior and were attracted by the strong dielectrophoretic forces generated by 3D carbon electrodes, the number of analyzed cells in Figure 2b is limited to 30 cells; however, the initial number of cells was always 3×10^6 cells/mL for the experiments (see Materials and Methods Section 2.2. Cell culture, Supplementary Video 2).

The monocyte population showed smooth nDEP (blue) to crossover (green) and crossover to pDEP (red) transition as a whole monocyte population, as shown in Figure 2a. On the other hand, the macrophage population exhibited more likely a bimodal distribution that is either the macrophage cells in nDEP (blue) or pDEP (red) in comparison to the monocyte population (Figure 2b). Therefore, the dielectric movement of the U937-differentiated macrophages showed more heterogeneous population responses than the U937 monocyte population which is the originals of U937-differentiated macrophages.

Figure 3 compares the dielectrophoretic movement of the U937 monocytes and U937-differentiated macrophages. The macrophages moved from the nDEP region to pDEP region when 50 kHz at 10 V_{pp} was applied. The monocytes experienced nDEP to pDEP transition when 100–150 kHz at 20 V_{pp} was provided. When both the monocyte and macrophage populations exhibited strong pDEP forces at 1 MHz, there was not any significant difference between the trapping regions of the cells according to Student's t-test (p value was 0.892, where * $p < 0.5$ was significant), as shown in Figure 3. This result may show that the interfacial polarization difference between the cytoplasm and plasma membrane can be stronger for macrophages than monocytes [54]. Therefore, the observed macrophage dielectric properties at 1 MHz can be related to both membrane and cytoplasm properties of macrophages, whereas the membrane features might dominate for the monocyte dielectric properties at 1 MHz. These varying biophysical properties between monocytes and macrophages might explain their distinct trapping regions inside the 3D carbon DEP device.

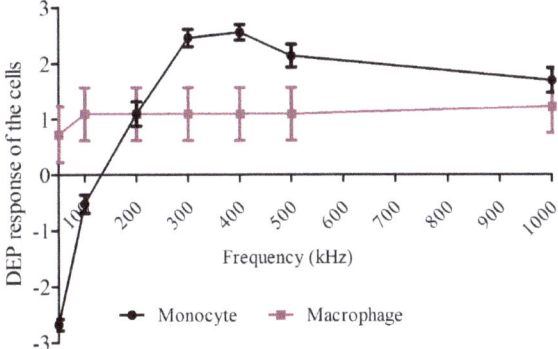

Figure 3. Comparison between the DEP movement of monocyte and macrophage cells. The magnitude of movement is categorized as very strong (3), strong (2), and weak forces (1). The "-" sign refers to nDEP. Measurements are the mean and error. n = 50 for monocytes, n = 30 for macrophages.

3.2. Dielectrophoretic Deformation Index

While dielectrophoretic forces distributed the cells in the electrode array according to their polarizability difference, DEP forces were also capable of creating deformation on the cells. As mentioned above, monocytes and macrophages are well-known cells for their plasticity properties [9,10]. When mammalian cells were exposed to large external flow forces in variable microenvironments using microfluidics, they became elongated, varied in size, and tended to return to their original shape once the external forces were removed [42,55].

We determined the dielectrophoretic deformation indexes (DDI) of the single U937 monocyte and the U937-differentiated macrophage cells using the non-uniform AC electric field varying from 50 kHz to 1 MHz frequency. The DDI values of each monocyte and macrophage cells were calculated for 47 cells as defined in Equation (4) [50], where H (μm) was the major and W (μm) was the minor axes of the cells, as shown in Figure 4a.

$$DI = \frac{H}{W} \quad (4)$$

Figure 4 illustrated the DDI distribution for the monocytes (Figure 4b,d) and macrophages (Figure 4c,e), including the outliers. Monocyte population demonstrated significant DDI difference between 0–400 kHz, and 50–400 kHz at 20 V_{pp} ($p < 0.5$, Section 2 Materials and Methods, Section 2.5 Image acquisition and data analysis). The increased pDEP forces made the monocytes taller while attracting to the strong pDEP regions. When the pDEP forces reaches their maximum, the monocyte cells became wider and their deformation index significantly decreased at 300 kHz–1 MHz, and 400 kHz–1 MHz, 20 V_{pp} ($p < 0.05$), as shown in Figure 4b. Monocyte cells tended to generate pearl chain like organization under the influences of strong pDEP forces. Figure 4d displays the underlying dynamics of monocyte population when the change of deformation index was followed for each single cell. Single-cell analysis was performed when the DEP forces were applied for 50–500 kHz. The deformation index for the U937 monocytes were dynamically changed and created a zig-zag pattern within the 0.433–2.147 boundaries. Contrary to the deformation of monocytes, macrophages did not considerably alter their deformation (Figure 4c). Figure 4e demonstrates the deformation index of single macrophages that was exposed to DEP forces for the frequency range of 50–500 kHz. The change of deformation index for the U937-differentiated macrophage cells was more stable than U937 monocytes. The deformation indexes of macrophages exhibited smooth trajectories within the boundaries of 0.457–1.588.

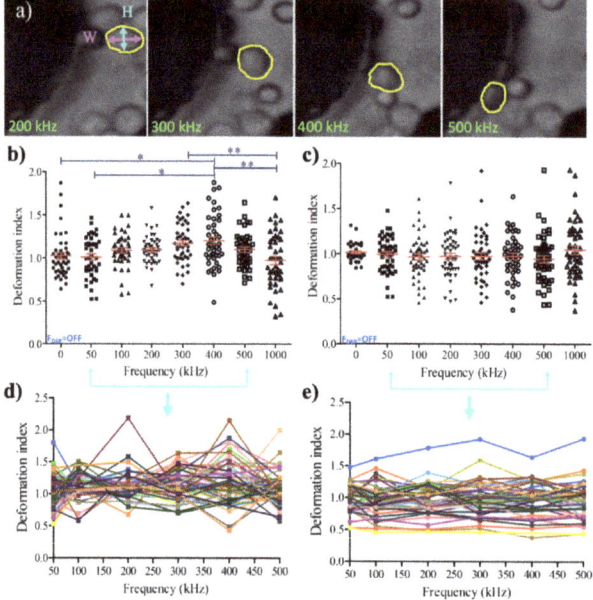

Figure 4. Dielectrophoretic deformation indexes (DDI) of U937 monocytes and U937-differentiated macrophages: (**a**) Representative image for the measurement of DDI. DDI values were presented with mean and standard error for population (n = 45). (**b**) single (n = 47) (**d**) monocyte cells; 45 population (**c**), single (n = 47) (**e**) macrophage cells. Tukey's multiple comparison test is applied for (**b**). * and ** indicate that data are significantly different with $p < 0.5$ and $p < 0.05$, respectively. Each color displays the change of deformation indexes of single cells during the frequencies applied for the range of 50–500 kHz in (**d**) and (**e**).

Figure 5 demonstrates that there was a significant DDI difference between U937 monocytes and U937-differentiated macrophages at 300 kHz ($p < 0.5$) and 400 kHz ($p < 0.05$) according to Tukey's multiple comparison test as explained in the Materials and Methods Section 2.5. (Image acquisition and data analysis). Monocyte population has higher DDI in comparison to macrophage population at 300 and 400 kHz, where both cell types were under the influences of pDEP forces. Next, increasing the frequencies decreased the DDI for monocyte cells, whereas it did not affect the DDI for macrophage cells.

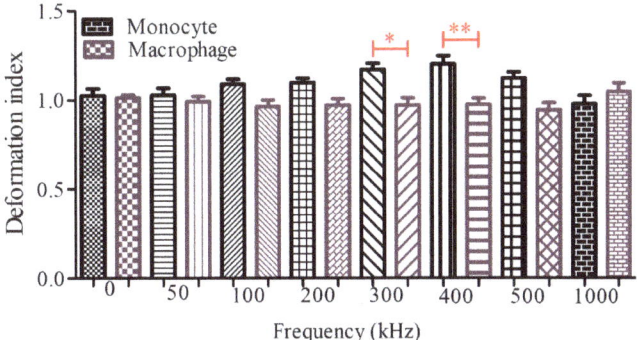

Figure 5. Comparison of the dielectrophoretic deformation indexes for the monocytes and macrophages without outliers. Measurements are the dielectrophoretic deformation index with mean and standard error for 45 monocyte and 45 macrophages cells. Tukey's multiple comparison test is applied. * and ** indicate that data are significantly different with $p < 0.5$ and $p < 0.05$, respectively.

3.3. Dielectric Mobility and Membrane Relaxation Time

The principle of examining the polarized particles with DEP has been implemented to reveal the biophysical properties of cells since 1962 [25,33,56–62]. The strongest motivation beyond these studies has been the development of label-free dielectric biomarkers to distinguish healthy and pathological cells, since surface charge density of cells plays key roles in exocytosis, endocytosis, cell adhesion [63,64], binding of proteins [65–67] etc. The electrophoretic behavior of single cells has been predicted using the mathematical models that define the relationship between the mobility and the surface charges acting upon a cell suspending in a low conductive medium [67].

Here, we investigated whether dielectric mobility (μ_{DEP}) [68] and membrane relaxation time (τ) [69] values are intrinsic, specific, dielectric markers that reliably distinguish U937 monocytes and U937-differentiated macrophages cell populations that have the same cell origin.

The dielectric mobility has been defined by Crowther and coworkers as in Equation (5), where η denotes the viscosity of the DEP buffer [68].

$$\vec{v}_{DEP} = -\mu_{DEP}\nabla|\vec{E}|^2 = -\left(\frac{\varepsilon_m r^2 K(w)}{3\eta}\right)\nabla|\vec{E}|^2 \tag{5}$$

The membrane relaxation time (τ) was expressed in Equation (6), where $C_{cell\ membrane}$ means the membrane capacitance of the cells [69].

$$\tau = rC_{cell\ membrane}\left(\frac{1}{\sigma_{cell\ membrane}} + \frac{1}{2\sigma_m}\right) \tag{6}$$

Using the equations above, the dielectrophoretic mobility and membrane relaxation time values were calculated with the physical and electrical properties of the monocyte and macrophage cells, and the low conductive DEP buffer, as presented in Table 1.

The dielectric mobilities were calculated as 6.99×10^{-18} m^4/V^{2}s and 12.40×10^{-18} m^4/V^{2}s for monocytes ($\mu_{DEP_{Monocyte}}$) and macrophages ($\mu_{DEP_{Macrophage}}$), respectively. The membrane relaxation time values for the monocytes ($\tau_{Monocyte}$) were 2.63×10^5 s, while ($\tau_{Macrophage}$) was 2.73×10^5 s for the macrophages. Here, the membrane capacitance values used for the calculations were not belonged to specifically for the U937 monocytes and U937-differentiated macrophages, as noted in Table 1 [70–72]. To the best of our knowledge, the exact membrane capacitance value for the U937 macrophages has not been yet measured. Therefore, the values in Table 1 should be carefully interpreted.

Table 1. Dielectric markers specific to U937 monocytes and U937-differentiated macrophages.

Parameters (Units)	Values	Resources
$r_{Monocyte}$ (m)	1.15×10^{-5}	Measured
$r_{Macrophage}$ (m)	1.5×10^{-5}	Measured
$K(\omega)_{Monocyte}$	0.976	Calculated [67]
$K(\omega)_{Macrophage}$	0.979	Calculated [67]
ε_m (C/V.m)	6.90×10^{-10}	-
η_{water} (kg/s.m)	8.90×10^{-4}	-
σ_m (S/m)	2×10^{-3}	Measured
$\sigma_{Monocyte\ membrane}$ (S/m)	7×10^{-13}	[70]
$\sigma_{Monocyte\ membrane}$ (S/m)	7×10^{-13}	[70]
$\sigma_{Macrophage\ membrane}$ (S/m)	7×10^{-13}	Assumed
$C_{Monocyte}$ (F/m^2)	0.016 ± 0.002	[70,71]
$C_{Macrophage}$ (F/m^2)	0.013 ± 0.001	[70,71]
$\tau_{Monocyte}$ (s)	2.63×10^5	Calculated
$\tau_{Macrophage}$ (s)	2.73×10^5	Calculated
$\mu_{DEP_{Monocyte}}$ (m^4/V^{2}s)	6.99×10^{-18}	Calculated
$\mu_{DEP_{Macrophage}}$ (m^4/V^{2}s)	12.40×10^{-18}	Calculated

4. Discussion

Monocytes and macrophage cells, sharing the same cell origin, have been compared according to their dielectrophoretic mobility and deformation. Both monocyte and macrophage populations exhibited inter-individual difference due to their intrinsic properties such as size, shape, and changes in membrane surface organization that may result in heterogeneity in their DEP responses.

Here, the crossover frequency of U937 monocytes was around 150 kHz. The U937-differentiated macrophage cells exhibited weak crossover frequency around 50 kHz (Figure 2). We used the computational tool for dielectric modeling published by Cottet, J. et al. and obtained the CM factor $K(\omega)$ values for the monocytes ($K(\omega)_{Monocyte}$) and macrophages ($K(\omega)_{Macrophage}$) as 0.976 and 0.979, respectively [67] (Table 1). Since the ($K(\omega)_{Macrophage}$) was slightly higher than the ($K(\omega)_{Monocyte}$), macrophages were exhibited pDEP behavior earlier than monocytes (see Figure 2b,c and Figure 3). The uniformity of pDEP responses of the monocytes was improved with increasing frequencies (see Supplementary Video 1), the macrophages displayed both nDEP and pDEP fractions for the whole frequencies ranging from 50 kHz–1 MHz (see Supplementary Video 2). Although there was no significant difference between the trapping regions of the cells (Student's t-test: p value was 0.892, where * $p < 0.5$ was significant), the DEP movement of macrophages were more heterogeneous than monocytes (Figure 3). We previously reported dielectrophoretic characterization and separation of U937 monocytes and U937-differentiated macrophages according to their crossover frequencies in [30–32].

This study, contrary to our previous work, reported that the translational DEP forces were not only moved cells according to their polarizability differences inside the electrode array, they also created irreversible deformation on the cells. Monocyte and macrophage cells display high plasticity among immune cells [9,10,42,55]. When DEP forces were introduced, the deformation index of monocytes first increased (0–400 kHz), then decreased with increasing pDEP forces (400 kHz–1 MHz), as shown in Figure 4. On the other hand, the deformation index of the macrophage cells did not exhibit significant difference for the frequencies ranging from 50 kHz to 1 MHz (Figure 4). When the dielectrophoretic deformation indexes of the monocyte and macrophage cell populations were compared, according to Tukey's multiple comparison test, the increase in the deformation index of monocytes was significantly higher than the deformation index of macrophages at 300 kHz ($p < 0.5$) and 400 kHz ($p < 0.05$), as shown in Figure 5. Here, we calculated the DEP deformation indexes of the cells (Figure 4a: location of the cells according to electrodes) as we measured their translational mobility due to applied F_{DEP} (Figure 2a: position of the cells according to electrodes). Therefore, it relied on the spatial distribution of the cells within the electrode array since the DEP forces depend on polarizability of the cells according to their intrinsic properties. The applied DEP forces synchronized the cells spatiotemporally within the electrode array and we measured the deformation of single cells at their specific locations when the specific frequencies and voltages were applied, therefore, we achieved to obtain consistent results for the dielectric deformation indexes of the cells (Figure 4d,e).

In addition to experimental results, the dielectrophoretic mobility and membrane relaxation time values were predicted using the physical and electrical properties of the monocyte and macrophage cells, and the low conductive DEP buffer (Table 1) [68–72]. The calculated values were quite similar both for monocytes and macrophage cells. However, the values in Table 1 should be carefully interpreted since there are still unknown dielectric parameters for the U937 monocytes and U937-differentiated macrophages.

Our results marshalled considerable evidence for the feasibility of using dielectric mobility and dielectric deformation index as a dielectric biomarker that presents biophysical differences between the cell lines which shares the same origin. To the best of our knowledge, this is the first study that presents dielectric deformation indexes of cells and may become a practical method for achieving a specific, high-throughput, continuous, label-free, sensitive electromechanical characterization and classification technique for U937 monocytes and U937-differentiated macrophages.

Our further studies will focus on separation and recovery of cells with different deformation indexes from the 3D carbon DEP platform for downstream analysis using immunostaining and quantitative reverse transcription-polymerase chain reaction (RT-qPCR) techniques. Hence, we can promptly explain the dielectrophoretic mobility and deformation differences in terms of transcription and protein expression levels in the membrane surface and cytoskeletal components. Moreover, we can employ this method for further characterization of macrophage subpopulations, and it may provide value in increasing our understanding of the nature of tumor associated macrophages (TAMs).

5. Conclusions

This study presents heterogeneity of monocytes and macrophages according to their intrinsic dielectrophoretic properties in terms of dielectrophoretic deformation indexes. We performed dielectric deformation measurements of the U937 monocytes and U937-differentiated macrophages with similar radius and dielectric characteristics using 3D carbon electrode microfluidic platform both at population- and single-cell level. We calculated deformation indexes of the cells when 10–20 V_{pp} voltage with frequencies ranging from 50 kHz to 1 MHz have been applied.

Our results showed that the crossover frequency for the monocytes was around 150 kHz [30–32]. Monocytes presented maximum deformation at 400 kHz and minimum deformation around 1 MHz frequencies at 20 V_{pp}. On the other hand, the crossover frequency for the macrophages, which differentiated from monocytes, was lower than 50 kHz, 10 V_{pp} [30–32]. Moreover, the dielectrophoretic deformation index for the macrophages was not significantly varied from 50 kHz to 1 MHz frequency range. We conclude that the change of the deformation index for macrophages was less in comparison to monocytes. Both dielectric mobility and deformation spectra revealed significant differences between the dielectric responses of U937 monocytes and U937-differentiated macrophages, which share the same origin.

Our method can be advanced for the development of label-free, specific, and sensitive single-cell characterization tools. This technique eliminates the possibility of damaging the cells by aggressive shear forces while allowing these cells to be used for further downstream analysis. To advance this work, we focus on development of automated image analysis tools to obtain directly deformation indexes and mobility data of cells from the acquired DEP videos.

Here, we particularly underlined F_{DEP}-generated deformation index of monocytes and macrophages, since these cells are among the white blood cells which are capable of infiltrating different types of tissues. Further DEP studies might interrogate to quantify other immune cells or their subsets (TAMs), and whether their intrinsic cellular heterogeneity can be quantified according to their dielectrophoretic deformation indexes.

Supplementary Materials: The following are available online at http://www.mdpi.com/2072-666X/11/6/576/s1, Video S1: Monocyte, Video S2: Macrophage.

Author Contributions: Conceptualization, M.E. and E.S.; methodology, M.E.; formal analysis, E.S.; investigation, M.E.; resources, M.E.; data curation, E.S.; writing—original draft preparation, E.S., M.E.; writing—review and editing, M.E.; visualization, M.E.; supervision, M.E.; project administration, M.E. All authors have read and agreed to the published version of the manuscript.

Funding: This research received no external funding.

Acknowledgments: We wish to thank Martinez-Duarte for providing the microfluidic chip and proofreading the article; Yagmur Yildizhan for her help for experimental data; Sumeyra Vural for valuable discussions.

Conflicts of Interest: The authors declare no conflict of interest.

References

1. Adekanmbi, E.O.; Srivastava, S.K. Dielectrophoretic applications for disease diagnostics using lab-on-a-chip platforms. *Lab Chip* **2016**, *16*, 2148–2167. [CrossRef]
2. Mulhall, H.J.; Labeed, F.H.; Kazmi, B.; Costea, D.E.; Hughes, M.P.; Lewis, M.P. Cancer, pre-cancer and normal oral cells distinguished by dielectrophoresis. *Anal. Bioanal. Chem.* **2011**, *401*, 2455–2463. [CrossRef]
3. Hur, S.C.; Henderson-Maclennan, N.K.; Mccabe, E.R.B.; Carlo, D.D. Deformability-based cell classification and enrichment using inertial microfluidics. *Lab Chip* **2011**, *11*, 912–920. [CrossRef]
4. Suresh, S.; Spatz, J.; Mills, J.; Micoulet, A.; Dao, M.; Lim, C.; Seufferlein, T. Connections between single-cell biomechanics and human disease states: Gastrointestinal cancer and malaria. *Acta Biomater.* **2005**, *1*, 15–30. [CrossRef]
5. Suresh, S. Biomechanics and biophysics of cancer cells. *Acta Biomater.* **2007**, *3*, 413–438. [CrossRef]
6. Pohl, H.A. The Motion and Precipitation of Suspensoids in Divergent Electric Fields. *J. Appl. Phys.* **1951**, *22*, 869–871. [CrossRef]
7. Pethig, R. Review Article—Dielectrophoresis: Status of the theory, technology, and applications. *Biomicrofluidics* **2010**, *4*, 022811. [CrossRef]
8. Cross, S.E.; Jin, Y.-S.; Rao, J.; Gimzewski, J.K. Nanomechanical analysis of cells from cancer patients. *Nat. Nanotechnol.* **2007**, *2*, 780–783. [CrossRef]
9. Hoffman, B.D.; Crocker, J.C. Cell Mechanics: Dissecting the Physical Responses of Cells to Force. *Ann. Rev. Biomed. Eng.* **2009**, *11*, 259–288. [CrossRef]
10. Chauviere, A.; Preziosi, L.; Byrne, H. A model of cell migration within the extracellular matrix based on a phenotypic switching mechanism. *Math. Med. Biol.* **2009**, *27*, 255–281. [CrossRef]
11. Taylor, P.R.; Gordon, S. Monocyte Heterogeneity and Innate Immunity. *Immunity* **2003**, *19*, 2–4. [CrossRef]
12. Hume, D.A. Plenary Perspective: The complexity of constitutive and inducible gene expression in mononuclear phagocytes. *J. Leukoc. Biol.* **2012**, *92*, 433–444. [CrossRef]
13. Lichtman, M.A.; Williams, W.J. Biochemistry and function of monocytes and macrophages. In *Williams Hematology*; McGraw-Hill Medical: New York, NY, USA, 2006; pp. 971–978.
14. Hulsmans, M.; Clauss, S.; Xiao, L.; Aguirre, A.D.; King, K.R.; Hanley, A.; Nahrendorf, M. Macrophages Facilitate Electrical Conduction in the Heart. *Cell* **2017**, *169*. [CrossRef]
15. Pittet, M.J.; Swirski, F.K. Monocytes link atherosclerosis and cancer. *Eur. J. Immunol.* **2011**, *41*, 2519–2522. [CrossRef]
16. Etzerodt, A.; Tsalkitzi, K.; Maniecki, M.; Damsky, W.; Delfini, M.; Baudoin, E.; Lawrence, T. Specific targeting of CD163 TAMs mobilizes inflammatory monocytes and promotes T cell–mediated tumor regression. *J. Exp. Med.* **2019**, *216*, 2394–2411. [CrossRef]
17. Thiriot, J.D.; Martinez-Martinez, Y.B.; Endsley, J.J.; Torres, A.G. Hacking the host: Exploitation of macrophage polarization by intracellular bacterial pathogens. *Pathog. Dis.* **2020**, *78*. [CrossRef]
18. Eming, S.A.; Krieg, T.; Davidson, J.M. Inflammation in Wound Repair: Molecular and Cellular Mechanisms. *J. Investig. Dermatol.* **2007**, *127*, 514–525. [CrossRef]

19. Swirski, F.K.; Nahrendorf, M.; Etzrodt, M.; Wildgruber, M.; Cortez-Retamozo, V.; Panizzi, P.; Pittet, M.J. Identification of Splenic Reservoir Monocytes and Their Deployment to Inflammatory Sites. *Science* **2009**, *325*, 612–616. [CrossRef]
20. Muller, P.A.; Koscsó, B.; Rajani, G.M.; Stevanovic, K.; Berres, M.-L.; Hashimoto, D.; Bogunovic, M. Crosstalk between Muscularis Macrophages and Enteric Neurons Regulates Gastrointestinal Motility. *Cell* **2014**, *158*, 300–313. [CrossRef]
21. Wynn, T.A.; Chawla, A.; Pollard, J.W. Macrophage biology in development, homeostasis and disease. *Nature* **2013**, *496*, 445–455. [CrossRef]
22. Epelman, S.; Lavine, K.J.; Randolph, G.J. Origin and Functions of Tissue Macrophages. *Immunity* **2014**, *41*, 21–35. [CrossRef]
23. Okabe, Y.; Medzhitov, R. Tissue-Specific Signals Control Reversible Program of Localization and Functional Polarization of Macrophages. *Cell J.* **2014**, *157*, 832–844. [CrossRef]
24. Xue, J.; Schmidt, S.V.; Sander, J.; Draffehn, A.; Krebs, W.; Quester, I.; Schultze, J.L. Transcriptome-Based Network Analysis Reveals a Spectrum Model of Human Macrophage Activation. *Immunity* **2014**, *40*, 274–288. [CrossRef]
25. Eisenberg, S.; Ben-Or, S.; Doljanski, F. Electro-kinetic properties of cells in growth processes. *Exp. Cell Res.* **1962**, *26*, 451–461. [CrossRef]
26. Petty, H.; Ware, B.; Liebes, L.; Pelle, E.; Silber, R. Electrophoretic mobility distributions distinguish hairy cells from other mononuclear blood cells and provide evidence for the heterogeneity of normal monocytes. *Blood* **1981**, *57*, 250–255. [CrossRef]
27. Bauer, J.; Hannig, K. Changes of the electrophoretic mobility of human monocytes are regulated by lymphocytes. *Electrophoresis* **1984**, *5*, 269–274. [CrossRef]
28. Yang, J.; Huang, Y.; Wang, X.; Wang, X.-B.; Becker, F.F.; Gascoyne, P.R. Dielectric Properties of Human Leukocyte Subpopulations Determined by Electrorotation as a Cell Separation Criterion. *Biophys. J.* **1999**, *76*, 3307–3314. [CrossRef]
29. Yang, J.; Huang, Y.; Wang, X.-B.; Becker, F.F.; Gascoyne, P.R. Differential Analysis of Human Leukocytes by Dielectrophoretic Field-Flow-Fractionation. *Biophys. J.* **2000**, *78*, 2680–2689. [CrossRef]
30. Elitas, M.; Yildizhan, Y.; Islam, M.; Martinez-Duarte, R.; Ozkazanc, D. Dielectrophoretic characterization and separation of monocytes and macrophages using 3D carbon-electrodes. *Electrophoresis* **2018**, *40*, 315–321. [CrossRef]
31. Yildizhan, Y.; Erdem, N.; Islam, M.; Martinez-Duarte, R.; Elitas, M. Dielectrophoretic Separation of Live and Dead Monocytes Using 3D Carbon-Electrodes. *Sensors* **2017**, *17*, 2691. [CrossRef]
32. Erdem, N.; Yildizhan, Y.; Elitas, M. A numerical approach for dielectrophoretic characterization and separation human hematopoietic cells. *Int. J. Eng. Res.* **2017**, *6*, 1079–1082. [CrossRef]
33. Tonin, M.; Bálint, S.; Mestres, P.; Martinez, I.A.; Petrov, D. Electrophoretic mobility of a growing cell studied by photonic force microscope. *Appl. Phys. Lett.* **2010**, *97*, 203704. [CrossRef]
34. Song, H.; Rosano, J.M.; Wang, Y.; Garson, C.J.; Prabhakarpandian, B.; Pant, K.; Lai, E. Continuous-flow sorting of stem cells and differentiation products based on dielectrophoresis. *Lab Chip* **2015**, *15*, 1320–1328. [CrossRef]
35. Salmanzadeh, A.; Kittur, H.; Sano, M.B.; Roberts, P.C.; Schmelz, E.M.; Davalos, R.V. Dielectrophoretic differentiation of mouse ovarian surface epithelial cells, macrophages, and fibroblasts using contactless dielectrophoresis. *Biomicrofluidics* **2012**, *6*, 024104. [CrossRef]
36. Polizzi, S.; Laperrousaz, B.; Perez-Reche, F.J.; Nicolini, F.E.; Satta, V.M.; Arneodo, A.; Argoul, F. A minimal rupture cascade model for living cell plasticity. *New J. Phys.* **2018**, *20*, 053057. [CrossRef]
37. Agrawal, R.; Smart, T.; Nobre-Cardoso, J.; Richards, C.; Bhatnagar, R.; Tufail, A.; Pavesio, C. Assessment of red blood cell deformability in type 2 diabetes mellitus and diabetic retinopathy by dual optical tweezers stretching technique. *Sci. Rep.* **2016**, *6*. [CrossRef]
38. Fabry, B.; Maksym, G.N.; Hubmayr, R.D.; Butler, J.P.; Fredberg, J.J. Implications of heterogeneous bead behavior on cell mechanical properties measured with magnetic twisting cytometry. *J. Magn. Magn. Mater.* **1999**, *194*, 120–125. [CrossRef]
39. Kim, Y.; Kim, K.; Park, Y. Measurement Techniques for Red Blood Cell Deformability: Recent Advances. *Blood Cell* **2012**. [CrossRef]

40. Popescu, G.; Ikeda, T.; Dasari, R.R.; Feld, M.S. Diffraction phase microscopy for quantifying cell structure and dynamics. *Opt. Lett.* **2006**, *31*, 775–777. [CrossRef]
41. Shin, S.; Ku, Y.-H.; Ho, J.-X.; Kim, Y.-K.; Suh, J.-S.; Singh, M. Progressive impairment of erythrocyte deformability as indicator of microangiopathy in type 2 diabetes mellitus. *Clin. Hemorheol. Microcirc.* **2007**, *36*, 253–261.
42. Zeng, N.F.; Mancuso, J.E.; Zivkovic, A.M.; Smilowitz, J.T.; Ristenpart, W.D. Red Blood Cells from Individuals with Abdominal Obesity or Metabolic Abnormalities Exhibit Less Deformability upon Entering a Constriction. *PLoS ONE* **2016**, *11*. [CrossRef] [PubMed]
43. Pinho, D.; Campo-Deaño, L.; Lima, R.; Pinho, F.T. In vitroparticulate analogue fluids for experimental studies of rheological and hemorheological behavior of glucose-rich RBC suspensions. *Biomicrofluidics* **2017**, *11*, 054105. [CrossRef] [PubMed]
44. Guido, I.; Xiong, C.; Jaeger, M.S.; Duschl, C. Microfluidic system for cell mechanics analysis through dielectrophoresis. *Microelectron. Eng.* **2012**, *97*, 379–382. [CrossRef]
45. Du, E.; Dao, M.; Suresh, S. Quantitative biomechanics of healthy and diseased human red blood cells using dielectrophoresis in a microfluidic system. *Extreme Mech. Lett.* **2014**, *1*, 35–41. [CrossRef]
46. Pires-Afonso, Y.; Niclou, S.P.; Michelucci, A. Revealing and Harnessing Tumour-Associated Microglia/Macrophage Heterogeneity in Glioblastoma. *Int. J. Mol. Sci.* **2020**, *21*, 689. [CrossRef]
47. Armistead, F.J.; Pablo, J.G.D.; Gadêlha, H.; Peyman, S.A.; Evans, S.D. Cells Under Stress: An Inertial-Shear Microfluidic Determination of Cell Behavior. *Biophys. J.* **2019**, *116*, 1127–1135. [CrossRef]
48. Martinez-Duarte, R. SU-8 Photolithography as a Toolbox for Carbon MEMS. *Micromachines* **2014**, *5*, 766–782. [CrossRef]
49. Martinez-Duarte, R.; Renaud, P.; Madou, M.J. A novel approach to dielectrophoresis using carbon electrodes. *Electrophoresis* **2011**. [CrossRef]
50. Martinez-Duarte, R.; Cito, S.; Collado-Arrendondo, E.; Martinez, S.O.; Madou, M.J. Fluid-dynamic and Electromagnetic Characterization of 3D Carbon Dielectrophoresis with Finite Element Analysis. *Sens. Transducers J.* **2008**, *3*, 25–36.
51. Natu, R.; Martinez-Duarte, R. Numerical Model of Streaming DEP for Stem Cell Sorting. *Micromachines* **2016**, *7*, 217. [CrossRef]
52. Nakano, A.; Ros, A. Protein dielectrophoresis: Advances, challenges, and applications. *Electrophoresis* **2013**, *34*, 1085–1096. [CrossRef] [PubMed]
53. Pohl, H.A.; Pollock, K.; Crane, J.S. Dielectrophoretic force: A comparison of theory and experiment. *J. Biol. Phys.* **1978**, *6*, 133–160. [CrossRef]
54. Huang, Y.; Wang, X.-B.; Becker, F.F.; Gascoyne, P.R. Membrane changes associated with the temperature-sensitive P85gag-mos-dependent transformation of rat kidney cells as determined by dielectrophoresis and electrorotation. *Biochim. Biophys. Acta* **1996**, *1282*, 76–84. [CrossRef]
55. Liang, W.; Liu, J.; Yang, X.; Zhang, Q.; Yang, W.; Zhang, H.; Liu, L. Microfluidic-based cancer cell separation using active and passive mechanisms. *Microfluid. Nanofluid.* **2020**, *24*. [CrossRef]
56. Vargas, F.F.; Osorio, M.H.; Ryan, U.S.; Jesus, M.D. Surface Charge of Endothelial Cells Estimated from Electrophoretic Mobility. *J. Membr. Biochem.* **1989**, *8*, 221–227. [CrossRef] [PubMed]
57. Ohshima, H.; Kondo, T. On the electrophoretic mobility of biological cells. *Biophys. Chem.* **1991**, *39*, 191–198. [CrossRef]
58. Wal, A.V.D.; Minor, M.; Norde, W.; Zehnder, A.J.B.; Lyklema, J. Electrokinetic Potential of Bacterial Cells. *Langmuir* **1997**, *13*, 165–171. [CrossRef]
59. Weiss, N.G.; Jones, P.V.; Mahanti, P.; Chen, K.P.; Taylor, T.J.; Hayes, M.A. Dielectrophoretic mobility determination in DC insulator-based dielectrophoresis. *Electrophoresis* **2011**. [CrossRef]
60. Harrison, H.; Lu, X.; Patel, S.; Thomas, C.; Todd, A.; Johnson, M.; Xuan, X. Electrokinetic preconcentration of particles and cells in microfluidic reservoirs. *Analyst* **2015**, *140*, 2869–2875. [CrossRef]
61. Cummings, E.B.; Singh, A.K. Dielectrophoresis in Microchips Containing Arrays of Insulating Posts: Theoretical and Experimental Results. *Analy. Chem.* **2003**, *75*, 4724–4731. [CrossRef]
62. Crowther, C.V.; Hayes, M.A. Refinement of insulator-based dielectrophoresis. *Analyst* **2017**, *142*, 1608–1618. [CrossRef]
63. Pillet, F.; Dague, E.; Ilić, J.P.; Ružić, I.; Rols, M.-P.; Denardis, N.I. Changes in nanomechanical properties and adhesion dynamics of algal cells during their growth. *Bioelectrochemistry* **2019**, *127*, 154–162. [CrossRef]

64. Ferrari, M.; Cirisano, F.; Morán, M.C. Mammalian Cell Behavior on Hydrophobic Substrates: Influence of Surface Properties. *Colloids Interfaces* **2019**, *3*, 48. [CrossRef]
65. Hess, B.; van der Vegt, N.F.A. Cation specific binding with protein surface charges. *Proc. Natl. Acad. Sci. USA* **2009**, *106*, 13296–13300. [CrossRef] [PubMed]
66. Birant, G.; Wild, J.D.; Meuris, M.; Poortmans, J.; Vermang, B. Dielectric-Based Rear Surface Passivation Approaches for Cu (In,Ga)Se2 Solar Cells—A Review. *Appl. Sci.* **2019**, *9*, 677. [CrossRef]
67. Cottet, J.; Fabregue, O.; Berger, C.; Buret, F.; Renaud, P.; Frénéa-Robin, M. MyDEP: A New Computational Tool for Dielectric Modeling of Particles and Cells. *Biophys. J.* **2019**, *116*, 12–18. [CrossRef]
68. Crowther, C.V.; Hilton, S.H.; Kemp, L.; Hayes, M.A. Isolation and identification of Listeria monocytogenes utilizing DC insulator-based dielectrophoresis. *Anal. Chim. Acta* **2019**, *1068*, 41–51. [CrossRef]
69. Punjiya, M.; Nejad, H.R.; Mathews, J.; Levin, M.; Sonkusale, S. A flow through device for simultaneous dielectrophoretic cell trapping and AC electroporation. *Sci. Rep.* **2019**, *9*. [CrossRef] [PubMed]
70. Holmes, D.; Pettigrew, D.; Reccius, C.H.; Gwyer, J.D.; Berkel, C.V.; Holloway, J.; Morgan, H. Leukocyte analysis and differentiation using high speed microfluidic single cell impedance cytometry. *Lab Chip* **2009**, *9*, 2881–2889. [CrossRef] [PubMed]
71. Sano, M.B.; Henslee, E.A.; Schmelz, E.; Davalos, R.V. Contactless dielectrophoretic spectroscopy: Examination of the dielectric properties of cells found in blood. *Electrophoresis* **2011**, *32*, 3164–3171. [CrossRef]
72. Khoshmanesh, K.; Nahavandi, S.; Baratchi, S.; Mitchell, A.; Kalantar-Zadeh, K. Dielectrophoretic platforms for bio-microfluidic systems. *Biosens. Bioelectron.* **2011**, *26*, 1800–1814. [CrossRef] [PubMed]

© 2020 by the authors. Licensee MDPI, Basel, Switzerland. This article is an open access article distributed under the terms and conditions of the Creative Commons Attribution (CC BY) license (http://creativecommons.org/licenses/by/4.0/).

Article

Real-Time Monitoring and Detection of Single-Cell Level Cytokine Secretion Using LSPR Technology

Chen Zhu [1], Xi Luo [1,2,*], Wilfred Villariza Espulgar [1], Shohei Koyama [3], Atsushi Kumanogoh [3], Masato Saito [1,2], Hyota Takamatsu [3] and Eiichi Tamiya [1,2,*]

1. Department of Applied Physics, Graduate School of Engineering, Osaka University, 2-1 Yamadaoka, Suita, Osaka 565-0871, Japan; chen@ap.eng.osaka-u.ac.jp (C.Z.); wilfred@ap.eng.osaka-u.ac.jp (W.V.E.); saitomasato@ap.eng.osaka-u.ac.jp (M.S.)
2. AIST-Osaka University Advanced Photonics and Biosensing Open Innovation Laboratory, AIST, 2-1 Yamadaoka, Suita, Osaka 565-0871, Japan
3. Immunology Frontier Research Center, Graduate School of Medicine, Osaka University, 2-2 Yamadaoka, Suita, Osaka 565-0871, Japan; koyama@imed3.med.osaka-u.ac.jp (S.K.); kumanogo@imed3.med.osaka-u.ac.jp (A.K.); thyota@imed3.med.osaka-u.ac.jp (H.T.)
* Correspondence: ra.luoxi@aist.go.jp (X.L.); tamiya@ap.eng.osaka-u.ac.jp (E.T.)

Received: 10 December 2019; Accepted: 15 January 2020; Published: 19 January 2020

Abstract: Cytokine secretion researches have been a main focus of studies among the scientists in the recent decades for its outstanding contribution to clinical diagnostics. Localized surface plasmon resonance (LSPR) technology is one of the conventional methods utilized to analyze these issues, as it could provide fast, label-free and real-time monitoring of biomolecule binding events. However, numerous LSPR-based biosensors in the past are usually utilized to monitor the average performance of cell groups rather than single cells. Meanwhile, the complicated sensor structures will lead to the fabrication and economic budget problems. Thus, in this paper, we report a simple synergistic integration of the cell trapping of microwell chip and gold-capped nanopillar-structured cyclo-olefin-polymer (COP) film for single cell level Interleukin 6 (IL-6) detection. Here, in-situ cytokine secreted from the trapped cell can be directly observed and analyzed through the peak red-shift in the transmittance spectrum. The fabricated device also shows the potential to conduct the real-time monitoring which would greatly help us identify the viability and biological variation of the tested single cell.

Keywords: localized surface plasmon resonance (LSPR) technology; Interleukin 6 (IL-6) detection; single cell trapping; single cell level immunoassay

1. Introduction

Cytokines are a broad and loose category of small immunological protein biomarkers secreted by the immune cells. They play a critical role in adjusting the cell signaling, cell differentiation and biological response in the human immune system, and are proven to be involved in cell autocrine, paracrine and endocrine signaling as immune-modulating agents [1–4]. Thus, the researches about cytokines have been a main focus of studies among the scientists in the recent decades. Among all the cytokines, IL-6 stands out for its outstanding contribution to clinical diagnosis and cell immunoassay. It is an interleukin that acts as both a pro-inflammatory cytokine, an anti-inflammatory myokine and also an important mediator of fever and acute phase responses [5]. In addition, the IL-6 is responsible for stimulating acute phase protein synthesis, as well as the production of neutrophils in the bone marrow. It supports the growth of B cells and is antagonistic to regulatory T cells [6–8]. Thus, the detection of IL-6 becomes our first target in this research.

The enzyme-linked immunosorbent assay (ELISA) is one of the most widely used conventional methods for cytokine detection recently. This conventional method allows sufficient quantification of target proteins via only a simple parallel array-type operation [9,10]. However, there still exists some weak points within this method. For example, the ELISA requires secondary antibodies that bind with target analytes and complex sample labeling, which make it time consuming [11]. To deal with the weak points of ELISA, the scientists report an improved technology named enzyme linked immunospot (ELISPOT) assay. The ELISPOT is a type of method that focuses on the quantitatively high-throughput measuring of single cell level cytokine secretion with much higher sensitivity [12]. Although numerous advantages could be provided by ELISPOT, there still remains a huge concern that the personal counting errors during the experiments will have an impact on the final results [13–15]. Another conventional method, fluorescence-based single cell intensity detection, requires multiple times fluorescence dyes staining which is also time consuming and complex. At the same time, a large amount of sample volume is needed, which has a great impact on the saving of precious samples especially in clinical applications [16–18]. Aside from the technologies mentioned above, the localized surface plasmon resonance (LSPR) is another widely used method for fast, label-free and real-time monitoring of biomolecule binding events [19–22]. The LSPR is a plasmonic phenomenon that arises around nanoscale structures or nanoparticles of noble metals when light is illuminated onto a nanoscale-featured sensing surface [21,23–25]. It will occur when the natural frequency of the oscillating conduction electrons of the conductive metal nanoparticles matches the incident light frequency, causing resonant oscillations of electrons [23,26,27]. Currently, LSPR-based biosensors utilize the biomolecular interactions that lead to the change of the refractive index (RI) in the vicinity of the sensing surface to conduct the spontaneous detection, which is proved highly significant for diagnostic and point-of-care testing (POCT) purposes [21,28]. Especially for detecting the antibody–antigen interactions, any changes of RI could result in a sensitive response in the LSPR-induced light absorption spectrum, which is beneficial for quantitative analysis. Until now, numerous LSPR-based biosensors integrated with microfluidics have been proposed for therapeutic applications and potential to realize the portable detecting platforms [29–31].

In order to fabricate the cost-effective LSPR-based biosensors, a cheap mass production technology such as nanoimprinting is extremely needed. Thus, in the previous research, we have already reported the fabrication of a nanoimprinted gold-capped nanopillar structures on cyclo-olefin-polymer (COP) for LSPR-based detections [4]. In this work, a microwell array and a gold-capped nanopillar structure were integrated into a simple analysis platform. Further research showed that microwell arrays could provide the relatively high single cell occupancy capability and meanwhile supply a suitable environment for long-time and real-time monitoring. Thousands of individual cells could be trapped and monitored within trap sites through simple gravitational sedimentation. The regular cell migration and cell-to-cell interactions could also be avoided by the trapping structure and well-to-well pitches [32–35]. In our proposed device, the edge area of the microwell structure was utilized to detect the RI change owing to the cytokine secretion and antibody binding.

In this research, fresh cultured IL-6 over-expressed Jurkat cells were utilized to evaluate the sensitivity and capability of our fabricated device. The cultured cells were directly trapped and started to release IL-6 which would immediately bind with the antibody on the surface of nanopillar-structured LSPR detection film without stimulation [6,36,37]. The result proved that our fabricated device has the potential to trap single cells reaching over 60% occupancy efficiency with relatively low cell concentrations and volumes, which is extremely significant in clinical diagnosis. Furthermore, the device shows the capability to detect the single cell transmittance spectrum peak red-shift caused by single cell cytokine secretions and a maximum of 1.8 nm peak shift was observed by our device through real-time cell monitoring.

2. Materials and Methods

2.1. Fabrication of Porous Alumina (PA) Mold

In this research, the nanoporous anodic alumina oxide (AAO) was first prepared as the mold for nanopillar structure formation on cyclo-olefin-polymer (COP) films. The whole fabrication process has already been reported in our previous report in detail [4]. Briefly, the AAO mold was fabricated using a two-step anodizing treatments. The first anodizing step was conducted under a constant voltage of 80 V for 1 h to generate an aluminum oxide layer on the polished aluminum plate. Afterwards, the layer was removed by sinking inside a mixer containing phosphoric acid (1.16%, w/v) and chromic acid (5%, w/v) at 60 °C. The second anodizing step was conducted under a constant voltage of 80 V for 1 min. Finally, the phosphoric acid etching was carried out for 13 min at 40 °C. Then, the treated mold was carefully dried by pure N_2 gas and stored for further use.

2.2. Fabrication and Immobilization of Gold-Capped Nanopillar Structured Polymer Film

The whole nanoimprinting procedure was conducted with X-300H (SCIVAX Corp., Kawasaki, Japan). A pressure of 0.83 MPa was applied for 1 min at 100 °C immediately after the PA-mold and COP (ZF-14-188) film were carefully arranged to the machine stage. Afterwards, the temperature was increased into 160 °C and the pressure was increased to 2 MPa for 10 min. Next, the pressure was released and the whole stage was cooling down to 80 °C. The processed COP film was carefully peeled off from the PA-mold. Next, the oxygen plasma etching was performed on the treated COP film to create an uneven structure on the surface of the nanopillar. According to the mechanism, it was considered that the COP resin surface was irradiated with oxygen (oxygen radicals) in a high energy state which would combined with carbon constituting the COP resin, vaporized and decomposed as CO_2 [38]. After 60 s oxygen plasma treatment, the surface of the pillar roughened as shown in the scanning electron microscope (SEM) image in Figure 1. The diameter of the larger size pillar ranged from 150–200 nm and the distance between pillars was ranged from 20–50 nm. In addition, due to the oxygen plasma treatment, several smaller pillars ranging from 30–50 nm were formed on the surface of larger pillars. Then, the transformed COP chip was sputtered with gold using the Compact Sputter machine (ULVAC ACS4000, Yokohama, Japan) to form a 35 nm layer of gold on the COP film surface. Afterwards, it came to the immobilization steps. First, the gold-sputtered COP chip was submerged into the 10 mL 10-carboxy-1decanethiol reagent for 30 min to from a self-assembled monolayer (SAM) layer on the COP film surface and carefully washed with 99.5% Ethanol followed by a drying step via pure N_2. Next, 100 µL of mixed reagent which consist of 0.1 M N-Hydroxysuccinimide (NHS) and 0.4 M 1-ethyl-3-(3-dimethylaminopropyl) carbodiimide hydrochloride (WSC) was uniformly dripped onto the surface of the COP film for 10 min and followed by a washing step via phosphate buffer saline (PBS) for activation. Furthermore, 100 µL of 50 ng/mL Anti-IL-6 was uniformly dripped onto the film surface for 30 min to combine with the previously made SAM layer. Finally, 100 µL of 1% BSA was coated onto the COP film surface for 30 min to block the whole structure followed by a washing step with PBST (PBS with 0.05% Tween-20) and PBS.

2.3. Fabrication of Cell Trapping Micro-Well Structured Chip

To fabricate the cell trapping chip, a clean silicon wafer mold was necessary. About 5 mL SU-8 3025 (Microchem, Newton, MA, USA) was first spin-coated on the surface of a clean 4-inch diameter silicon wafer according to the data sheet provided by the manufacturer at around 3000 rpm for 30 s to form a 20 µm uniform SU-8 layer. Afterwards, the cured SU-8 layer was exposed under the previously prepared glass mask and ultraviolet (UV) light for 6 s using a mask aligner (Mikasa, MA-10, Tokyo, Japan). The exposed mold was then treated under post exposure bake (PEB) protocol at 95 °C for 4 min followed by a developing process. Finally, the mold was rinsed, dried and stored in a clean container for further use. A clean 2 mm thick COP chip was carefully covered on the surface of the fabricated silicon mold and transferred into the nanoimprint machine. The nanoimprint step was performed under 1000 N pressure for 10 min under 40 °C. Afterwards, the COP chip was carefully removed from the mold and the cell trapping chip was successfully fabricated with the well diameter ranging from 13–15 μm. When the diameter of the microwell was set to 10 μm or lower, almost no single cells could be trapped by our device. Besides, if the diameter of the microwell was set to 16 μm or higher, there would be a high possibility to have cells stacking problems. Thus, here in our research, we choose to use the 13 μm as the standard diameter of the microwells.

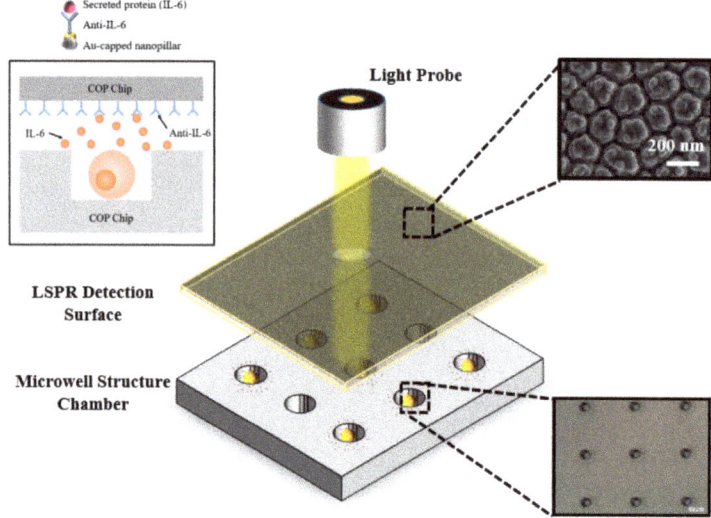

Figure 1. Schematic of the integrated localized surface plasmon resonance (LSPR) cytokine detection platform device.

2.4. Sensitivity Evaluation of the Gold-Capped COP Chip

The evaluation of the gold-capped COP film sensitivity is a quite significant index for analyzing the utility of the chip. In our research, the transmittance spectrum of the gold-capped COP surface was measured with the microscope (OLYMPUS IX70) with a spectrometer. Different refractive index environments were first evaluated, such as H_2O (n = 1.33), 1 M glucose (n = 1.35), ethylene glycol (n = 1.43) and glycerol (n = 1.47). The absorption spectrum peak red-shift which resulted by the LSPR phenomenon was observed and plotted. The slope of the curve (peak shift/refractive index) was determined as the bulk sensitivity of the fabricated COP film. In addition, different concentrations of the IL-6 reagents were measured as the positive group, and the IgA reagent was measured as the negative group. Followed by this protocol, the limitation of detection (LOD) for IL-6 could be calculated.

2.5. Single Cell Occupancy Capability Evaluation of the Trapping Chip

The fresh cultured Jurkat cells were utilized to measure the occupancy capability of the fabricated micro-well structure COP chip. The cell trapping chip was first surrounded by the silicon rubber sheet for better injection of cell suspensions. Next, the chip was washed with MilliQ water and vacuumed to remove the trapped bubbles. Afterwards, 100 μL of cultured Jurkat cell suspension (concentration is 1×10^5 cells/mL) was carefully pipetted onto the surface of the chip and waited for 15 min for gravity sedimentation. After cell sedimentation, a syringe was utilized to provide suitable fluid power, which would help to pipe out the extra media and cells from the device to avoid the cell stacking problems, as demonstrated in Figure S2. Finally, 10 μL of CD31 conjugated with FITC was utilized for fluorescence staining and the chip was stored under 4 °C for 30 min before monitoring. Counting the single cell position using this fluorescence-based technique could also greatly reduce the possibility to treat the stacked cells as the single cells. Optical observation was performed directly using an inverted microscope (Olympus IX-71, Tokyo, Japan) with ×10 magnification.

2.6. Real-time Monitoring of Single Cell IL-6 Secretion Situation

Fresh cultured IL-6 over-expressed Jurkat cells which could release a relatively greater number of IL-6 cytokines used for real-time cell monitoring. All the experimental tools were washed and autoclaved in advance to avoid any contamination. The cells were first cultured in the common media: 89% Roswell Park Memorial Institute (RPMI) 1640 media, 10% Fetal Bovine Serum (FBS) and 1% penicillin. Next, the cultured cells were carefully dripped onto the surface of the trapping chip and we waited 15 min for sedimentation. The COP detection film was cut into 1×1 cm^2 pieces before the LSPR measurement. Afterwards, the cut film was covered on the top of the trapping chip, bound tightly and the integrated device was carefully placed under the microscope (OLYMPUS IX70, Tokyo, Japan) for real-time observation. The cell spectrum data was recorded every 6 min until 54 min, and the peak shifts in the absorption peak wavelength were recorded and visualized in a bar graph.

3. Results and Discussion

3.1. Morphology Characterization of the COP Detection and Trapping Device

The whole procedure to fabricate the single cell cytokine secretion detection device is shown in Figure 1. The device is simply the combination between the nanopillar-structured COP detection film and thick COP cell trapping chip, which is easily fabricated and portable. Figure 1 also shows the scanning electron microscope (SEM) image of the pillar structure. The anodization conditions have already been optimized for ease of removing the nanoimprinted COP film from the mold and formation of the nanostructures. In this research, several different well diameters and depths were evaluated to find the optimized conditions. As the diameter of target Jurkat cells were ranging from 8–12 µm, the single cell occupancy efficiency became relatively low when the trapping well diameter was less than 10 µm or over 16 µm which would lead to a high possibility to trap none or multiple cells. Meanwhile, the well depth also has a high impact on the cell occupancy efficiency. The experiment showed that the well depth ranging from 15–20 µm was more suitable for cell trapping. Thus, in our research, we finally chose to use the 13 µm diameter and 19 µm depth as the standard data to fabricate the trapping chip. Figure 1 (bottom) demonstrates the whole cell trapping chip and micro-well structures. In the whole design, 5000 single cell trapping wells are fabricated within the 1×1 cm^2 area for high-throughput research and the well pitch is adjusted to 100 µm to avoid any cell-to-cell contaminations.

3.2. Sensitivity Evaluation of the COP Film via Transmittance Spectrum Peak Red-Shift

The LSPR transmittance peak shifts data of the fabricated COP detection films were recorded in several different refractive index reagents. Different reagents were uniformly dripped onto the surface of the COP films separately, and the peak shift data was illustrated in Figure 2.

According to the spectrum data, with the RI increases, the transmittance spectrum has a larger peak red-shift, which indicates that the fabricated COP detection film exhibits working plasmonic properties. Meanwhile, the bulk sensitivity of the COP detection film, the transmittance spectrum peak shifts are plotted in Figure 2, which shows a slope of 190.2 nm/RIU (Refractive index unit). The sensitivity results clearly claim that the fabricated COP detection film has higher sensitivity compared to the previously reported LSPR film [4], and could respond to the changes in RI with corresponding transmittance peak red-shifts, which is highly significant in LSPR-based single cell detection.

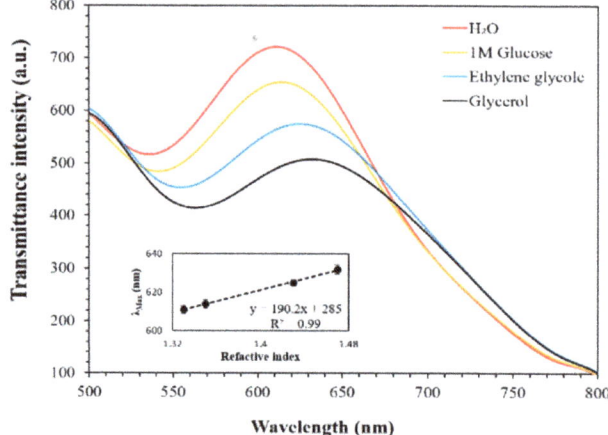

Figure 2. Transmittance LSPR peak wavelength shows red-shift response as observed in H_2O (n = 1.33), 1 M glucose (n = 1.35), ethylene glycol (n = 1.43) and glycerol (n = 1.47) and individual wavelength shifts over the refractive index reveal the average sensitivity of the fabricated plasmonic device.

3.3. IL-6 Calibration Curve Detection Based on COP Detection Film

As IL-6 is our target cytokine in this research, it is necessary to evaluate the calibration curve for further detection. 100 µL of 10 ng/mL, 25 ng/mL, 50 ng/mL, 100 ng/mL and 200 ng/mL concentrations of IL-6 reagents are separately dripped onto the surface of the immobilized COP detection film for 30 min followed by a washing step. Afterwards, the COP detection films are observed under the microscope and the transmittance spectrum peak shifts are recorded and plotted.

Figure 3 shows the calibration curve obtained from the responses recorded by each different target concentrations. The result shows that the fabricated COP detection film has a linear response with a detection limitation of 10 ng/mL.

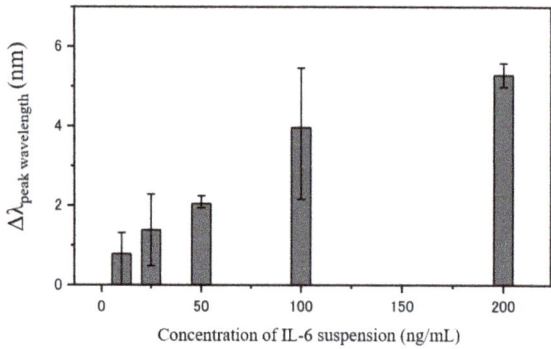

Figure 3. Calibration curve for concentrations 10 ng/mL, 25 ng/mL, 50 ng/mL, 100 ng/mL and 200 ng/mL IL-6 peak transmittance using fabricated plasmonic device.

3.4. Single Cell Occupancy Performance and Real-Time Monitoring of Single Cell Cytokine Secretion

The single cell occupancy efficiency of the COP trapping chip was evaluated using the Jurkat cells (φ = 8–12 µm). In the experiment, 100 µL of the cell suspension (concentration is 1×10^5 cells/mL) was dripped onto the surface of the trapping chip and we waited for 15 min for cell gravity sedimentation. Figure 4 shows the bright field image and also the fluorescence image of the trapped cells to clearly demonstrate that the cells are successfully isolated and trapped inside the micro-well structures. According to the observation result, our fabricated micro-well trapping devices could reach almost 60% single cell occupancy efficiency, as shown in Table 1.

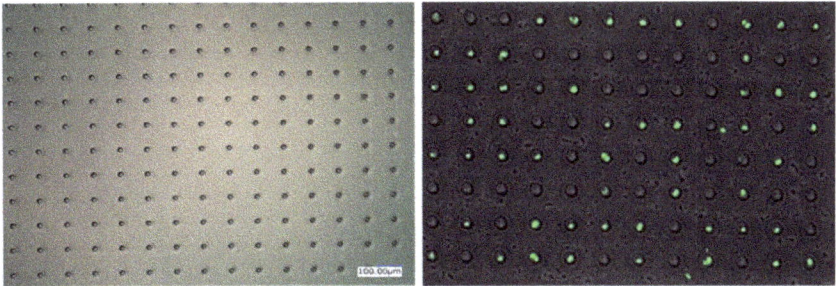

Figure 4. Evaluation of single cell occupancy efficiency using the fluorescence-based detection.

Table 1. Evaluation of different occupancy efficiency.

Cell Occupancy Type	Single Cell	Double Cells	Multiple Cells
Occupancy Efficiency	~60%	~10%	~5%

Real-time monitoring of healthy single IL-6 over-expressed Jurkat cells were conducted for 54 min until the cells were dead. LSPR spectrum peak shift measurement of the single cell was recorded every 6 min and the data was illustrated in the Figure 5a. However, due to the limitation of our imaging system, it would take 30–40 s for single cell imaging scan and ~10 single cells were scanned at one time. This result to a scan limit of every 6 min and in the future, we will improve the techniques to conduct the single cell scan in a shorter time to realize the high-throughput analysis. In general, the transmittance peak tended to red-shift with time passing and remained the same until the final scan as the cells were already dead. The maximum transmittance spectrum peak red shift was 1.8 nm. At the same time, the control group experiment was conducted only using the new cell culture media. The result in Figure 5b shows that with the time passing, the cell culture media do not have great impact on the transmittance spectrum, which proves that the peak shift changes are resulted by the binding of IL-6 secreted by the single cell and immobilized anti-IL-6. Besides, we replaced the anti-IL-6 with anti-IgA during immobilization step as another negative control group, and found that there was also no obvious peak shift in the peak wavelength before and after binding, as demonstrated in Figure S1, which proved that the peak wavelength shift was caused by the specific binding of secreted IL-6 and anti-IL-6. This result indicates that our LSPR detection device has the potential to analyze the single cell level cytokine secretion situations.

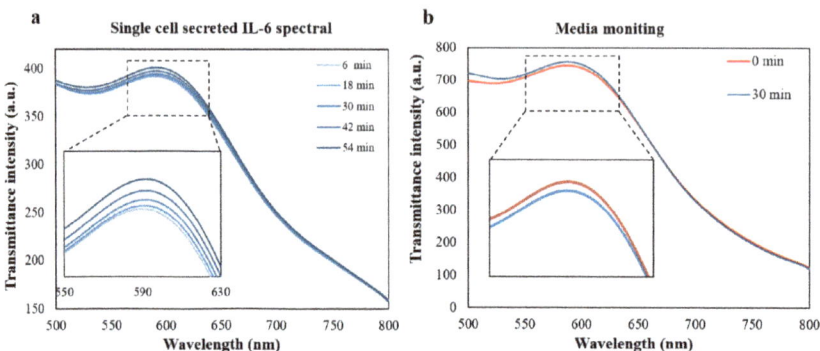

Figure 5. (**a**) Real-time transmittance spectrum observation of fresh-cultured IL-6 overexpressed single Jurkat cell using the fabricated plasmonic device during 54 min. (**b**) Negative control group to evaluate the effect of cell culture media to the LSPR transmittance spectrum detection after 30 min culturing.

4. Conclusions

In this research, a simple and portable LSPR detection device for single cell level cytokine secretion research is fabricated. With this device, over 3000 single cells could be isolated and trapped within the trapping sites at one time which is highly significant for the saving of precious samples especially in clinical medical diagnosis. In summary, our fabricated LSPR detection device could reach almost 60% single cell occupancy efficiency with relatively low sample concentrations and volumes. The bulk sensitivity of the device is found to be 190.2 nm/RIU which is proved capable for LSPR detection. A detection limitation of 10 ng/mL for anti-IL 6 is established using the fabricated film. Furthermore, real-time monitoring of healthy IL-6 over-expressed Jurkat cells are conducted on the device which shows a maximum of 1.8 nm peak red-shift during the one-hour detection period. The limitation of this study is not easy to realize the change of cell culture media for longer time monitoring. In the future, we will focus upon the improvement of the single cell trapping device which could help realize the longer time monitoring and also conduct the real patient sample experiments using the peripheral blood mononuclear cell (PBMC).

Supplementary Materials: The following are available online at http://www.mdpi.com/2072-666X/11/1/107/s1, Figure S1. Negative control group of anti-IgA to confirm the peak wavelength shift before and after binding. Figure S2. Cell trapping and washing procedures. After cell sedimentation, we use the syringe to provide fluid power which could help to pipe out the extra media and cells from the device to increase the single cell occupancy efficiency. Figure S3. Single cell peak wavelength shift corresponding to the time. Figure S4. Calibration curve for concentrations 1 ng/mL, 10 ng/mL, 25 ng/mL, 100 ng/mL, 200 ng/mL, 500 ng/mL, 750 ng/mL and 1000 ng/mL IL-6 peak wavelength using fabricated plasmonic device in dry conditions (dry with pure N_2 gas).

Author Contributions: C.Z., X.L., W.V.E., M.S. and E.T. designed the research. C.Z., X.L., S.K., H.T. and A.K. prepared the sample and performed the experiments. E.T. and A.K. reviewed and edited the manuscript. All the authors analyzed the data and discussed the results. All authors have read and agreed to the published version of the manuscript.

Funding: This work was supported in part by a Grant-in-Aid for Scientific Research (Kiban S, No. 15H05769) and JST CREST Grant number JPMJCR16G2, Japan.

Conflicts of Interest: The authors declare no conflict of interest.

References

1. Benton, H.P. Cytokines and their receptors. *Curr. Opin. Cell Biol.* **1991**, *3*, 171–175. [CrossRef]
2. Martins e Silva, J. [Biochemical characterization and metabolic effects of tumor necrosis factor]. Caracterizacao bioquimica e efeitos metabolicos do factor de necrose tumoral. *Acta Med. Port.* **1991**, *4* (Suppl. 1), 20S–27S.

3. Oh, B.R.; Huang, N.T.; Chen, W.Q.; Seo, J.H.; Chen, P.Y.; Cornell, T.T.; Shanley, T.P.; Fu, J.P.; Kurabayashi, K. Integrated nanoplasmonic sensing for cellular functional immunoanalysis using human blood. *ACS Nano* **2014**, *8*, 2667–2676. [CrossRef] [PubMed]
4. Saito, M.; Kitamura, A.; Murahashi, M.; Yamanaka, K.; Hoa, L.Q.; Yamaguchi, Y.; Tamiya, E. Novel gold-capped nanopillars imprinted on a polymer film for highly sensitive plasmonic biosensing. *Anal. Chem.* **2012**, *84*, 5494–5500. [CrossRef] [PubMed]
5. Fishman, D.; Faulds, G.; Jeffery, R.; Mohamed-Ali, V.; Yudkin, J.S.; Humphries, S.; Woo, P. The effect of novel polymorphisms in the interleukin-6 (IL-6) gene on IL-6 transcription and plasma IL-6 levels, and an association with systemic-onset juvenile chronic arthritis. *J. Clin. Investig.* **1998**, *102*, 1369–1376. [CrossRef] [PubMed]
6. Akira, S.; Hirano, T.; Taga, T.; Kishimoto, T. Biology of multifunctional cytokines—Il-6 and related molecules (Il-1 and Tnf). *FASEB J.* **1990**, *4*, 2860–2867. [CrossRef]
7. Barton, B.E. IL-6: Insights into novel biological activities. *Clin. Immunol. Immunopathol.* **1997**, *85*, 16–20. [CrossRef]
8. Schindler, R.; Mancilla, J.; Endres, S.; Ghorbani, R.; Clark, S.C.; Dinarello, C.A. Correlations and interactions in the production of Interleukin-6 (Il-6), Il-1, and tumor necrosis factor (Tnf) in human-blood mononuclear-cells—Il-6 suppresses Il-1 and Tnf. *Blood* **1990**, *75*, 40–47. [CrossRef]
9. Buss, H.; Chan, T.P.; Sluis, K.B.; Domigan, N.M.; Winterbourn, C.C. Protein carbonyl measurement by a sensitive ELISA method (vol 23, pg 361, 1997). *Free Radic. Biol. Med.* **1998**, *24*, 1352.
10. Voller, A.; Bartlett, A.; Bidwell, D.E. Enzyme immunoassays with special reference to Elisa techniques. *J. Clin. Pathol.* **1978**, *31*, 507–520. [CrossRef]
11. Lequin, R.M. Enzyme immunoassay (EIA)/enzyme-linked immunosorbent assay (ELISA). *Clin. Chem.* **2005**, *51*, 2415–2418. [CrossRef] [PubMed]
12. Patiris, P.; Hanson, C. Single-round HIV type 1 neutralization measured by ELISPOT technique in primary human cells. *AIDS Res. Hum. Retrovir.* **2005**, *21*, 784–790. [CrossRef] [PubMed]
13. Fujihashi, K.; Mcghee, J.R.; Beagley, K.W.; Mcpherson, D.T.; Mcpherson, S.A.; Huang, C.M.; Kiyono, H. Cytokine-specific elispot assay—Single cell analysis of Il-2, Il-4 and Il-6 producing cells. *J. Immunol. Methods* **1993**, *160*, 181–189. [CrossRef]
14. Okamoto, Y.; Gotoh, Y.; Tokui, H.; Mizuno, A.; Kobayashi, Y.; Nishida, M. Characterization of the cytokine network at a single cell level in mice with collagen-induced arthritis using a dual color ELISPOT assay. *J. Interferon Cytokine Res.* **2000**, *20*, 55–61. [CrossRef]
15. Zand, M.S.; Henn, A.D.; Brown, M.A.; Rebhan, J.; Murphy, A.J.; Coca, M.N.; Hyrien, O.; Mosmann, T. Measurement of single-cell IgG secretion rates by quantitative ELISPOT (qELISPOT): Modulation by BCR-crosslinking and cell division in CpG stimulated IgG+CD27+ human memory B cells. *Am. J. Transplant.* **2009**, *9*, 471.
16. Cohen, D.; Dickerson, J.A.; Whitmore, C.D.; Turner, E.H.; Palcic, M.M.; Hindsgaul, O.; Dovichi, N.J. Chemical cytometry: Fluorescence-based single-cell analysis. *Annu. Rev. Anal. Chem.* **2008**, *1*, 165–190. [CrossRef]
17. Mark, H.F.L.; Rehan, J.; Mark, S.; Santoro, K.; Zolnierz, K. Fluorescence in situ hybridization analysis of single-cell trisomies for determination of clonality. *Cancer Genet. Cytogenet.* **1998**, *102*, 1–5. [CrossRef]
18. Wang, Y.Z.; DelRosso, N.V.; Vaidyanathan, T.V.; Cahill, M.K.; Reitman, M.E.; Pittolo, S.; Mi, X.L.; Yu, G.Q.; Poskanzer, K.E. Accurate quantification of astrocyte and neurotransmitter fluorescence dynamics for single-cell and population-level physiology. *Nat. Neurosci.* **2019**, *22*, 1936–1944. [CrossRef]
19. Hiep, H.M.; Nakayama, T.; Saito, M.; Yamamura, S.; Takamura, Y.; Tamiya, E. A microfluidic chip based on localized surface plasmon resonance for real-time monitoring of antigen-antibody reactions. *Jpn. J. Appl. Phys.* **2008**, *47*, 1337–1341. [CrossRef]
20. Park, J.H.; Byun, J.Y.; Mun, H.; Shim, W.B.; Shin, Y.B.; Li, T.; Kim, M.G. A regeneratable, label-free, localized surface plasmon resonance (LSPR) aptasensor for the detection of ochratoxin A. *Biosens. Bioelectron.* **2014**, *59*, 321–327. [CrossRef]
21. Raphael, M.P.; Christodoulides, J.A.; Delehanty, J.B.; Long, J.P.; Pehrsson, P.E.; Byers, J.M. Quantitative LSPR imaging for biosensing with single nanostructure resolution. *Biophys. J.* **2013**, *104*, 30–36. [CrossRef]
22. Sepulveda, B.; Angelome, P.C.; Lechuga, L.M.; Liz-Marzan, L.M. LSPR-based nanobiosensors. *Nano Today* **2009**, *4*, 244–251. [CrossRef]

23. Bellapadrona, G.; Tesler, A.B.; Grunstein, D.; Hossain, L.H.; Kikkeri, R.; Seeberger, P.H.; Vaskevich, A.; Rubinstein, I. Optimization of localized surface plasmon resonance transducers for studying carbohydrate-protein interactions. *Anal. Chem.* **2012**, *84*, 232–240. [CrossRef] [PubMed]
24. Blaber, M.G.; Henry, A.I.; Bingham, J.M.; Schatz, G.C.; Van Duyne, R.P. LSPR imaging of silver triangular nanoprisms: Correlating scattering with structure using electrodynamics for plasmon lifetime analysis. *J. Phys. Chem. C* **2012**, *116*, 393–403. [CrossRef]
25. Zhang, Z.Y.; Chen, Z.P.; Qu, C.L.; Chen, L.X. Highly sensitive visual detection of copper ions based on the shape-dependent LSPR spectroscopy of gold nanorods. *Langmuir* **2014**, *30*, 3625–3630. [CrossRef] [PubMed]
26. Huang, T.; Nallathamby, P.D.; Xu, X.H.N. Photostable single-molecule nanoparticle optical biosensors for real-time sensing of single cytokine molecules and their binding reactions. *J. Am. Chem. Soc.* **2008**, *130*, 17095–17105. [CrossRef]
27. Mayer, K.M.; Lee, S.; Liao, H.; Rostro, B.C.; Fuentes, A.; Scully, P.T.; Nehl, C.L.; Hafner, J.H. A label-free immunoassay based upon localized surface plasmon resonance of gold nanorods. *ACS Nano* **2008**, *2*, 687–692. [CrossRef]
28. Wang, X.L.; Cui, Y.; Irudayaraj, J. Single-cell quantification of cytosine modifications by hyperspectral dark-field imaging. *ACS Nano* **2015**, *9*, 11924–11932. [CrossRef]
29. Fujiwara, K.; Watarai, H.; Itoh, H.; Nakahama, E.; Ogawa, N. Measurement of antibody binding to protein immobilized on gold nanoparticles by localized surface plasmon spectroscopy. *Anal. Bioanal. Chem.* **2006**, *386*, 639–644. [CrossRef]
30. Guo, L.H.; Kim, D.H. LSPR biomolecular assay with high sensitivity induced by aptamer-antigen-antibody sandwich complex. *Biosens. Bioelectron.* **2012**, *31*, 567–570. [CrossRef]
31. Kim, D.K.; Park, T.J.; Tamiya, E.; Lee, S.Y. Label-free detection of leptin antibody-antigen interaction by using LSPR-based optical biosensor. *J. Nanosci. Nanotechnol.* **2011**, *11*, 4188–4193. [CrossRef] [PubMed]
32. Kim, S.; Hall, E.; Zare, R.N. Microfluidics-based cell culture for single-cell analysis. In *Biophysical Journal*; Biophysical Society: Rockville, MD, USA, 2007.
33. Mazutis, L.; Gilbert, J.; Ung, W.L.; Weitz, D.A.; Griffiths, A.D.; Heyman, J.A. Single-cell analysis and sorting using droplet-based microfluidics. *Nat. Protoc.* **2013**, *8*, 870–891. [CrossRef] [PubMed]
34. Thompson, A.M.; Paguirigan, A.L.; Kreutz, J.E.; Radich, J.P.; Chiu, D.T. Microfluidics for single-cell genetic analysis. *Lab Chip* **2014**, *14*, 3135–3142. [CrossRef] [PubMed]
35. Yin, H.B.; Marshall, D. Microfluidics for single cell analysis. *Curr. Opin. Biotechnol.* **2012**, *23*, 110–119. [CrossRef] [PubMed]
36. Shimada, M.; Andoh, A.; Hata, K.; Tasaki, K.; Araki, Y.; Fujiyama, Y.; Samba, T. IL-6 secretion by human pancreatic periacinar myofibroblasts in response to inflammatory mediators. *J. Immunol.* **2002**, *168*, 861–868. [CrossRef]
37. Vgontzas, A.N.; Bixler, E.O.; Lin, H.M.; Prolo, P.; Trakada, G.; Chrousos, G.P. IL-6 and its circadian secretion in humans. *Neuroimmunomodulation* **2005**, *12*, 131–140. [CrossRef]
38. Hwang, S.J.; Tseng, M.C.; Shu, J.R.; Yu, H.H. Surface modification of cyclic olefin copolymer substrate by oxygen plasma treatment. *Surf. Coat. Technol.* **2008**, *202*, 3669–3674. [CrossRef]

© 2020 by the authors. Licensee MDPI, Basel, Switzerland. This article is an open access article distributed under the terms and conditions of the Creative Commons Attribution (CC BY) license (http://creativecommons.org/licenses/by/4.0/).

MDPI
St. Alban-Anlage 66
4052 Basel
Switzerland
Tel. +41 61 683 77 34
Fax +41 61 302 89 18
www.mdpi.com

Micromachines Editorial Office
E-mail: micromachines@mdpi.com
www.mdpi.com/journal/micromachines

www.ingramcontent.com/pod-product-compliance
Lightning Source LLC
LaVergne TN
LVHW070622100526
838202LV00012B/705